Algarve Plants and Landscape

Vila do Bispo with date-palm

Algarve Plants and Landscape

Passing Tradition and Ecological Change

D. J. MABBERLEY
Wadham College and Department of Plant Sciences
University of Oxford

AND

P. J. PLACITO
Wadham College
University of Oxford

With photographs by the authors
and drawings by Rosemary Wise

OXFORD UNIVERSITY PRESS

1993

Oxford University Press, Walton Street, Oxford OX2 6DP

Oxford   New York   Toronto
Delhi   Bombay   Calcutta   Madras   Karachi
Kuala Lumpur   Singapore   Hong Kong   Tokyo
Nairobi   Dar es Salaam   Cape Town
Melbourne   Auckland
and associated companies in
Berlin   Ibadan

Oxford is a trade mark of Oxford University Press

Published in the United States
by Oxford University Press Inc., New York

A catalogue record for this book is available from the British Library

Library of Congress Cataloging in Publication Data
Mabberley, D. J.
Algarve plants and landscapes: passing tradition and ecological
change / D. J. Mabberley and P. J. Placito; with photographs by the
authors and drawings by Rosemary Wise.
Includes bibliographical references and index.
1. Botany—Portugal—Algarve—Ecology. 2. Plants, Cultivated-
-Portugal—Algarve. 3. Agricultural ecology—Portugal—Algarve.
4. Phytogeography—Portugal—Algarve. 5. Botany—Portugal—Algarve-
-Pictorial works. 6. Plants, Cultivated—Portugal—Algarve-
-Pictorial works. 7. Landscape—Portugal—Algarve. 8. Algarve
(Portugal)—History. I. Placito, P. J. II. Title.
QK330.M33   1993   581.9469'6—dc20   92-26904
ISBN 0-19-858702-3

Typeset by
Expo Holdings, Malaysia.
Printed in Hong Kong

Foreword

José Manuel Teixera Gomes Pearce de Azevedo, OBE
(Lic. em Cien. Econ. e Financeiras)
HM British Consul, Portimão

'Algarve plants and landscape', o título de uma verdadeira Enciclopédia Algarvia. São oito Capítulos, uma tarefa trabalhosa e sugestiva, em que de uma maneira notável dois Professores da Universidade de Oxford, Senhores Dr D. J. Mabberley e P. J. Placito, dão vida a um extraordinário trabalho sobre a Província do Algarve, fazendo a todo o momento evidenciar os seus charme e pitoresco.

Através dêsse trabalho, ficarão os leitores a conhecer o Algarve sob os seus múltiplos aspectos quer eles sejam o aspecto histórico, humano, geográfico, climatérico, ecológico, cultural, quer seja a sua agricultura tradicional, o seu artesanato, etc.

Permito-me contudo salientar o realce que os autores da obra dão às plantas em geral, bem como às belas flores, convidando os leitores a visitar o Algarve a apreciá-las em determinadas épocas do ano.

Se no trabalho em questão, sem dúvida do maior interêsse para os estudiosos, sobressai a ciência dos autores, por outro, nota-se claramente a mão e o espírito de alguém (P.J.P.), que muito amor tem dedicado a esta bela região de Portugal, sua verdadeira segunda Pátria, onde reside há cerca de 20 anos.

Muito me honrou o convite para ser o autor do Preâmbulo de uma obra de tão elevado nível que em muito servirá a Cultura Portuguesa, e da qual sem dúvida beneficiarão residentes e visitantes que melhor queiram conhecer esta Província de tão histórico passado, e que todos os Algarvios desejam ver cada vez mais próspera, a bem do futuro de Portugal.

Aos Drs Mabberley e Placito, os meus parabéns. Como Algarvio e como ex-Presidente do Turismo da Província, o meu muito e sincero obrigado pela honra que decidiram conceder-me.

'Algarve plants and landscape' is the title of a real encyclopaedia of the Algarve. Here are eight chapters making up a significant, painstaking book by D.J. Mabberley and P.J. Placito of Oxford University. It is an extraordinary work on the picturesque charm of the Algarve.

Throughout the book, the reader is made aware of many aspects of the province: its history, geography, climate, ecology, people, culture, traditional agriculture, crafts, and so on.

I would stress the importance which the authors attach to the plant-life in general as well as to the more striking flowers, which will certainly entice the reader to visit the Algarve to appreciate them in their respective flowering seasons.

It is clear that the knowledge of the authors, one of whom (P.J.P.) has been living in the Algarve as his second home for 20 years, will be of great interest to students who will see that the book was obviously written with a great love for this beautiful region of Portugal.

I am honoured by the invitation to write the Foreword to a work of such a high standard, which will undoubtedly benefit both residents and visitors who wish to learn something of this province, which has such an important history and which all Algarvians wish to see prosper with the future of Portugal.

To David Mabberley and Peter Placito, I offer my congratulations. As an Algarvian and a former President of Tourism of the province, I give sincere thanks for the honour afforded me.

Preface

This book is written for the Algarve's many residents and visitors who want to know something of the plants, both wild and cultivated, in Portugal's southernmost province, and how these plants relate to the varied landscape. Our approach, taking into account the agriculture and other land uses of the area, the system by which so much of the characteristic scenery is maintained, as well as the native plants in both town and country, is a new one. The landscape has been modified by humans for so many thousands of years that any attempt to deal with the plants of the Algarve without such a consideration is, in our opinion, misguided.

We have written the book as a continuous narrative, and have illustrated it not only with photographs and line drawings of plants (almost all drawn from living rather than from museum material), but also with reproductions of nineteenth century aquatints — again, we believe, a novel feature. The thread running through the book is to integrate the physical background of the countryside in terms of geology and climate with hydrology and human geography. We take in the ecology of the region, zone by zone, from maritime to montane, including farming and other rural industries as they occur, and end with a consideration of urban plantings. The theme is rounded off with an attempt to set the Algarve as it is today in an historical context, with some thoughts on its future.

Notwithstanding the continuous thread, individual chapters can be read in isolation so that, for instance, those most interested in the wild plants would go straight to Chapters 2, 3, and 6, while those concerned more with agricultural practice and land use would use Chapters 4 and 5, and those anxious to find information about plants around hotels and in towns would turn to Chapter 7.

Throughout we have attempted to deal with that which is most apparent and therefore has an effect on the landscape; many rare plants have had to be excluded and we have not attempted to be botanically exhaustive. However, information about particular plant species may be found in a variety of places in the book. For example, those readers interested in identifying a particular plant would in the first instance turn to the illustrations and information printed near them in Chapters 2–7, but use of the index will lead them to other parts of the text where more information is presented. For example a conspicuous but often unfamiliar tree to visitors is the carob (Figures 138 and 139). This is discussed in detail (on pages 122–5), though further information on its importance in the vegetation classification of the Algarve is to be found on page 220, its introduction by the Moslems on page 228, and its changing position in the local economy on page 247.

The Algarve has been undergoing a period of rapid change since 1970, and especially since the early 1980s. These changes are closely linked with the expansion of tourism, and also with Portugal's recent entry into the European Community (EC). The EC has made available very substantial structural funds for projects concerned with tourism, road and other communications, agriculture, and so on. Thus it is that the face of the Algarve has altered in a major way, and is continuing to do so. Indeed so fast have been these changes, with many of them carried out without environmental impact studies having been made at all or otherwise effectively taken into consideration, that much of the Algarve's past richness of flora has been destroyed, and much more endangered. Indeed it is not going too far to say that the changes resulting from the enforced decay of the small farms have already led to far-ranging social effects, and certainly also to large-scale changes in the successful integrated agricultural methods which have been so characteristic of the Algarve for hundreds of years. We feel impelled to write this book while adequate information may be gathered on which to base it, and while what is left of this great legacy is still apparent.

We believe that our book will be of interest to visitors and residents throughout the year, but will stimulate those particularly interested in finding many wild plants in flower to visit the Algarve in March and April, or in May and June for the montane flora. We have therefore concentrated on those plants likely to be in flower at those times. We hope that all will find something that they did not know about the province of the Algarve, and that visitors will have their holidays made more interesting by reading the text. We also hope that the beauty and botanical richness, part of the heritage of Portugal, can be assured not only for future generations of the Portuguese people and other Europeans, but for all the peoples of the world.

Oxford   D. J. M.
January 1993   P. J. P.

Acknowledgments

The information used in this book has been collected since 1973. Much of the practical data has been derived from the Field Courses led by DJM for undergraduates from the Department of Plant Sciences, University of Oxford. We are indebted to these successive classes and to colleagues from Oxford, notably Quentin Cronk, Roger Hall, Caroline Pannell, Jenny Steele, Fred Topliffe, and Stan Woodell, but especially to Rosemary Wise who prepared the line illustrations. We are grateful for continued support and encouragement from Professor F. R. Whatley, FRS, and Professor C. J. Leaver, FRS, FRSE. We thank Anne Sing and Ann Stedman who have done so much work behind the scenes in Oxford not only for this but for many other books.

In Portugal, we thank especially Professor Eng° Fernando Abecassis (Fundação Luso-Americana, Lisbon), Professor Dr M. M. B. Amaral Fortes (Instituto Superior Técnico, Lisbon), Eng° David Assoreira (Presidente, Comissão Coordenação Região do Algarve, Faro), Coronel Barroso da Silva (Serviço Cartográfico do Exercito, Lisbon), Dr Manuel Bravo de Lima (Estação Agronómica Nacional, Oeiras), Horácio Cavaco Guerreiro (Presidente, Região de Turismo do Algarve), Dr Alvaro Cavaleiro (Ipocork Lda.), Centro Nacional de Protecção de Productos Agrícola (Oeiras), Howard Elton (Lisbon), D^{ra} Maria Dalila Espírito Santo (Instituto Superior de Agronomia, Lisboa), Peter Gay (Oporto), Dr Rui Goncalves Henriques (Centro Nacional Investigação Geográfica, Lisboa), Instituto Geográfico e Cadastral (Faro and Lisbon), Instituto Nacional do Ambiente (Lisbon), Instituto Nacional de Estatística (Lisbon), Instituto Nacional de Meteorologica e Geofísica (Lisbon), Prof. D^{ra} Teresa Judice Gamito (Universidade do Algarve), Eng^{a} Carlota Lagoa (Liga para a Protecção da Natureza, Lisbon), Eng°. Filomeno Machado (Serviço Nacional de Parques, Reservas e Conservação da Natureza, SNPRCN: Parque Nacional da Ria Formosa), Professor J. Malato Beliz (Universidade de Évora), Dr Joaquim Marques Ferreira (SNPRCN, Lisbon), Dr Armando Marques Guedes (Universidade de Lisboa, Lisbon), Maria Antónieta Mauro Martins (Núcleo Extensão Rural de Portimão), José Alberto Monteiro (Lisbon), D^{ra} Maria Eugénia S. A. Moreira (Centro de Estudos Geográficos, Universidade de Lisboa), D^{ra} Manuela d'Oliveira (British Council, Lisbon), Dr. Luis Palma (Universidade do Algarve), Eng^{a} Helena Gomes (Area de Paisagem Protegida do Sudoeste Alentejano e Costa Vicentina, Aljezur), Dr Armando Reis Moura and D^{ra} Graça Silva (Reserva da Ria Formosa), Li and Rolf Rohrlapper (Portimão), Serviços Geológicos de Portugal, Eng° Agr° E. L. da Silva Ferreira, Direcção Regional de Agricultura do Algarve (Faro, Patacão), Dr José Manuel Teixera Gomes Pearce de Azevedo, OBE (Portimão), and Arq° José Vasconcelos (SNPRCN, Lisbon). We appreciate the valuable help given to us by the Librarians of the Instituto Nacional Investigação de Científica (Lisbon), Instituto Superior de Agronomia (Lisbon), Laboratório Nacional de Engenharia Civil (Lisbon), Biblioteca Municipal de Portimão, Arquivos de Portimão, Arquivos de Faro, and of the Fundacão Calouste Gulbenkian (Lagos).

Much of the draft text was completed in the spring of 1990 at Portimão, where Margherita Christine Placito provided comforts necessary to sustain the authors, and

António Rodriques Serrenho continued to explain the traditional Algarvian land practices. We are indebted also to Henry Noltie of the Royal Botanic Garden Edinburgh, who was under the impression that he was visiting us for a holiday, but whose knowledge of the flora of Andalucia was invaluable in the fieldwork as well as for assistance with herbarium specimens.

We thank Dr Jonathan Edwards (Bournemouth University), Peter Goldblatt (Missouri Botanical Garden, USA), Peter Green and Brian Mathew (Royal Botanic Gardens, Kew), Charlie Jarvis of the British Museum (Natural History), Professor Vincent May (Bournemouth University), the Director of the Royal Horticultural Society (Wisley), and Max Walters, all of whom helped us over particular detailed problems. We thank M. -H. Cornaert (Commission of the European Communities, Brussels, Belgium), as well as Franz Ledl and Ernestine Rüscher in Austria. We acknowledge with gratitude the assistance provided by the Librarians of the Department of Plant Sciences, the Department of Geography, the Department of Zoology, the Ashmolean Museum, and the Bodleian Library, all in Oxford. Major Donald Baker kindly lent the pollinating insects illustrated in Rosemary Wise's drawings. Dr Steve Simpson (Curator, University Museum, Oxford) has helped us in the identification of the insect species mentioned throughout.

Jack Suhl (Alcalar) read Chapters 4 and 5 and substantially improved the text; Andrew Henley-Welch (Sociedade Agrícola Viveiros do Foral Limitada, São Bartolomeu de Messines, Algarve) provided many valuable comments on Chapter 5. Robert Pullan (University of Liverpool) assisted us greatly with Chapters 1 and 2.

Dr Hanneke Wirtjes (Wadham College, Oxford) read the entire manuscript and made many helpful suggestions.

Additional photographic material has been provided by courtesy of David J. Tucker (Wadham College, Oxford), Patrick Garton (St. Catherine's College, Oxford), Dr Eduardo G. Crespo, and D[ra] Maria E. Oliveira as well as by the Warden of the A Rocha Observatory, Mexilhoeira Grande. Our especial thanks are due to Paul and Madeline Allen-Luckman (Travelpress Europe, Ltd) who most generously gave access to their comprehensive photographic archives and permitted us to use a number of slides. Foto Tempera (Portimão) provided valued technical assistance in the production of many of the coloured illustrations.

We acknowledge with gratitude the assistance given to us in the preparation of this book by the Oxford University Press.

Contents

Plates

Plates 1–208 are between pages 48 and 49; Plates 209–279 are between pages 216 and 217.

1

Geography and ecology

The Algarve lies between parallels of latitude 37° 35′ and 36° 58′ North, and between longitude 7° 25′ and 9° 00′ West. It is the southernmost province of Portugal and occupies 5411 km². For comparison, the total area of continental Portugal, including the river estuaries of the Rivers Tagus (*Tejo*) and Sado is 89 060 km², which is approximately a seventh of the Iberian Peninsula. The Algarve forms an oblong strip averaging 135 km in length from its eastern boundary with Spain marked by the Guadiana river to the western margin at the Atlantic, and with a width of 30–40 km. Its northern boundary is the Serra de Espinhaço de Cão and the Serra de Monchique to the west, with the Serra do Caldeirão and Monte Figo to the east; this boundary is an artificial one since the highly folded *serras* extend northwards into Alentejo province. The southern boundary of the Algarve is the Atlantic.

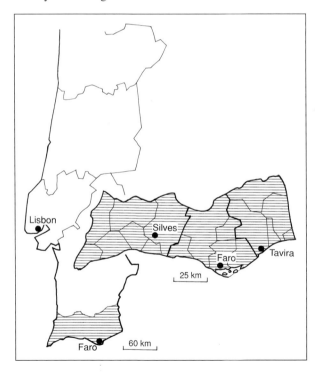

Figure 1 The Algarve as a province of Portugal with *concelho* divisions. *O Algarve no contexto regional do pais.* (Redrawn from Comissão de Coordenação da Região do Algarve 1985, Anexo A 1.)

To the east, divided from it by the Guadiana river, is the Spanish province of Huelva. The population of the Algarve (1990) is about 375 000 (1981 census, 323 534). The proportion of the Portuguese population in the province, 3.5 per cent, is precisely the same as in the 1639 census. In the Algarve the population is concentrated in the *Litoral*, the coastal zone, with a density of about 167 per km². This declines to 51 per km² in the *Barrocal*, the east–west strip of land between the southern coast and the hills and mountains to the north, and 18 per km² in the *Serra*, the northernmost part of the province. The average for the entire province is 65 inhabitants per km².

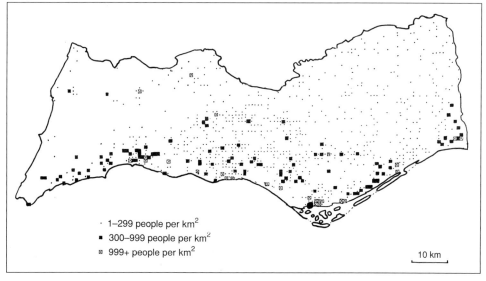

Figure 2   The Algarve: population densities. *Estrutura de povoamento, densidade.* (Redrawn from Comissão de Coordenação da Região do Algarve 1985, Anexo B 2.1.)

History

In the first centuries AD, the kingdom of the Algarve, at the time called *Conii* by the Romans, comprised the maritime strip of territory from Cape St. Vincent (*Cabo de S. Vicente*) to Almeria and incorporated other coastal Andalucian cities, these now being part of Spain, as well as parts of North Africa. Some authors hold that the word Algarve is derived from the Arabic *al-Gharb al-Andalus* (*o Ocidente* (the west) of Andalucia); this word is used in documents dating from 1189. The term *al-Gharb* is a synonym of *al-Moghreb*, today's *Almagrebe*. There are three *Almagrebes*, which are *A. Alacça* now Morocco, *A. Alauçate* now Algeria, and *A. Aladna* consisting of parts of Tunisia and Tripolitania–Libya. The area of *al-Gharb* east of the Guadiana river became Andalucia (*al-Andalus*), while that west of the river became known as *al-Faghar* (*Xenchir*).

From the reign of King (*Dom*) Sancho I (1185–1211), official papers refer to 'the King of Portugal of Silves and of Algarve'. This followed the subjection of the Moslem capital town of Silves (*Xelb, Chelb*) in 1189. The will (*testamento*) of Dom Dinis (1299) divided Portugal into six zones, with what is now the Algarve included

Figure 3   Silves Castle with citrus orchards.

within *Além Tejo e Odiana*. An ordinance (*Lei*) of 30 August 1406, defined the 'kingdom' of the Algarve as the southernmost province. The designation of kingdom persisted until 1599. Finally, the region now known as the Algarve, although with boundaries close to those of 1406, was termed a district (*distrito*) of Portugal. By the Decree of 18 July 1835, this became the *Distrito de Faro*, that is to say, a province of Portugal, with the city of Faro as its capital.

Long before that, Faro was a major port. Reports of 1811 refer to exports of dried fruits (figs), cork, sumach, and of rush baskets. Bananas and dates were grown there for local consumption and market boats were taking vegetables and other agricultural produce to Cadiz for example. Portimão, although the Algarve's second town today, was commercially insignificant, in 1811 its principal export being dried figs. At that time nearby Tavira was an important port exporting olive oil, dried figs, locust (carob), and sumach. Today the District of Faro is divided into the 16 councils (*concelhos*) of Albufeira, Alcoutim, Aljezur, Alportel, Castro Marim, Faro, Lagoa, Lagos, Loulé, Monchique, Olhão, Portimão, Silves, Tavira, Vila do Bispo, and Vila Real de Santo António. In turn these Councils are further sub-divided into 73 parishes (*freguesias*). Although the Algarve is officially a *distrito* (province), it is noteworthy that the title of King of Algarve was still included within the style of the last King of Portugal, Dom Manuel II, who in 1910 was exiled to England, where he died in 1932.

Why was it that the Algarve retained its separate identity as a kingdom until quite recent times? The Lusitanian people of the region had been overrun, although they were rarely under serious pressure, by successive waves of Phoenicians, Greeks, Carthaginians, Romans (from about 200 BC), and Moslems, generally referred to as

Figure 4   Portimão Harbour.

Figure 5   Alvôr.

Figure 6   Castro Marim.

Figure 7   Praia da Rocha.

Figure 8   The *Serra* from the *Barrocal* near Alte.

Figure 9   Alcoutim.

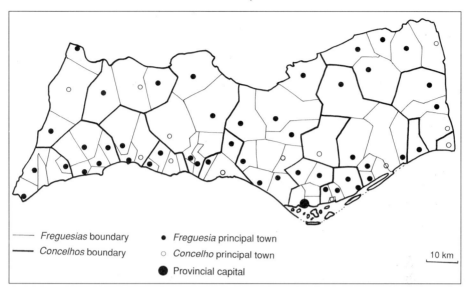

Figure 10   The Algarve: administrative divisions. *Divisão administrativa.*
(Redrawn from Comissão de Coordenação da Região do Algarve 1985, Anexo A 2.)

'Moors' (the effect of all these peoples on the agricultural development of the Algarve
is discussed in Chapter 8). So it was that a specific and distinct character developed in
the Algarve, quite different from those provinces which now form the remainder of
mainland Portugal. Moreover, the mountain ranges (*serras*) divide the province from

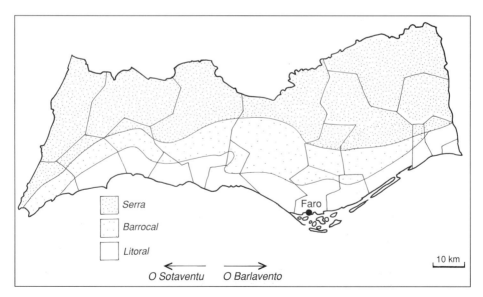

Figure 11   The Algarve: main characteristics of the province and its natural
divisions — *divisões naturais* — into the *Serra*, *Barrocal*, and *Litoral*.
(Redrawn from Comissão de Coordenação da Região do Algarve 1985, Anexo A 3.)

its northern neighbour Alentejo, and the Algarve was self-contained in the sense that its industrious people occupied a fertile land supporting almond, carob, fig, and olive trees and grapevines, with a coastline notable for plentiful fish and salt industries.

Furthermore, its ports, and especially Sagres, were strategically important. They were not only exporting valuable agricultural produce but were the bases for sea expeditions to Africa and India which, inspired by Prince Henry the Navigator (*O Infante*), led from 1415 onwards to the conquest of Ceuta and to the great 'Discoveries' (*Os Descobrimentos*) throughout the fifteenth century.

Geography

The Algarve falls geographically into three distinct zones which can be further divided. Their geology is an important factor in explaining the distribution of wild plants and governs the land-use régimes (see below).

1. To the south lies the coastal zone (*Litoral*). With an altitude up to 50 m, it makes up 1181 km², 21.8 per cent of the province. From the point of view of vegetation types, the *Litoral* can be considered as the 'west *Litoral*', the coastal strip north of Cape St. Vincent, and the 'south *Litoral*', the coastal strip along the south coast from Cape St. Vincent to Vila Real de Santo António.

2. To the north of the *Litoral* is the *Barrocal*. This zone has a prominent north-facing northern boundary with a gentle dipslope surface. Of its area some 1250 km² lie at an altitude of 100–300 m, representing a further 23 per cent of the area of the province.

3. The mountainous zone (*Serra*) is the northern part of the province. Some 7.3 per cent of the area of the Algarve, 390 km² has an altitude above 400 m. The major

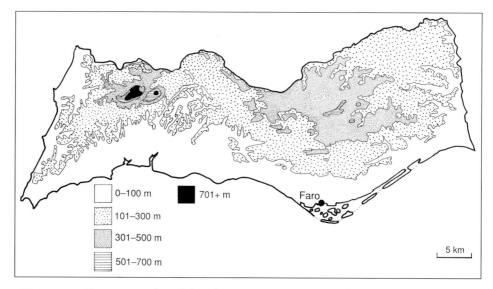

Figure 12 The topography of the Algarve. *Hipsometria.* (Redrawn from Comissão de Coordenação da Região do Algarve, 1985, Anexo A 4.)

part is the distinctive zone known as the Serra de Monchique, which has the peak of Foia (902 m) as the highest point in the province. Six km to the east on the same range is the peak of Picota (773 m), the second highest peak. Separating the Serra de Monchique from the Serra do Caldeirão is the depression between São Marcos de Serra and Santa Bartolomeu de Messines. The Serra do Caldeirão has a number of peaks exceeding 400 m, with the highest (Pelados) at 589 m. Within the *Serra* region, the area around Monchique itself can be considered as a separate sub-zone. The presence of good soils, and the availability of plenty of water for irrigation throughout the year, mainly from underground springs, combined with terracing, has made possible intensive cultivation. A further sub-zone within the *Serra* is the *Planalto*, an elevated plain in the Serra do Caldeirão.

Project Corine: biotopes and land cover

A remote-sensing imaging programme, known as Programa Corine (Co-ORdinated INformation on the Environment in the European Community), using space satellites, and covering the Algarve, was initiated in 1976. On 27 June 1985, the Council of the European Communities adopted this programme for gathering and distributing information which included features of the environment and natural resources in the Algarve. These surveys were carried out by the Landsat 2 and, later, 'second-generation' Landsat 4 satellites using the multispectral scanning system (MSS) and

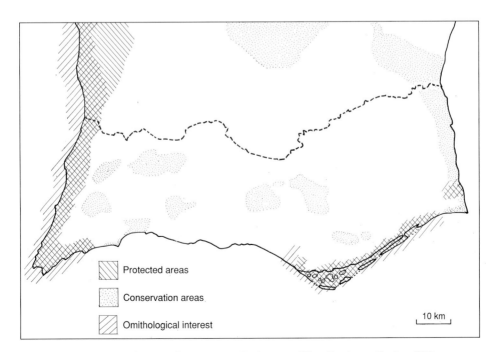

Figure 13 Areas of scientific interest, Corine satellite. *Projecto Corine-Biótopos.* (Redrawn from Anon. 1989*f*.)

thematic mapper (TM) imaging, the latter having considerably higher resolution. Various methods of digital data processing have been used, for example contrast enhancements, principal component analysis, and multispectral rectangular classification. In each case, the data so far derived have provided a detailed classification of areas such as 'wetlands', largely salt-marshes and mud-flats, and barrier islands including dunes, creeks, tidal deltas, and river inlets. Areas of the Algarve delineated in this way include the Reserve of Ria Formosa, the Bay of Lagos, and the confluence of rivers bounding the Alvôr peninsula. While there are many difficulties in interpreting the data obtained, detailed information regarding soil erosion and land cover is being derived from the *Corine* programme, as is a classification of the many types of soil in the province. Overall, the results from *Corine* have been disappointing, especially when they have not been supported by expert groundwork.

Climate

The Mediterranean region lies between the land mass of northern Europe and the deep belt of desert which is North Africa. The characteristic climatic feature is the combination of mild wet winters and long dry summers. From April to October there is normally little rain, apart from the occasional shower. However, winters are quite different from those of more northerly areas of Europe and even heavy rain is often followed by sun, the winter rains tending to be concentrated downpours rather than steady showers. One consequence of this is that considerable soil erosion takes place, especially on hillsides. Moreover, coastal erosion in the Algarve, with waves typically ranging from 2 m up to 6 m, and often associated with nearby building works and removal of plant cover, is a much more significant factor than in the 'closed' inland Mediterranean areas of Morocco, Spain, and Italy.

In general, climatic conditions in the Algarve have been reasonably constant for the past 8500 years, that is since the end of the last major Ice Age. Climatically the province is broadly divisible into two parts. The *Barlavento* (meaning windward), the westernmost part, is subject to the Atlantic and the substantial wind forces arising there — with prevailing winds from the north and north-west. The other is the *Sotavento* (leeward), the eastern part of the Algarve which extends to the border with Spain. There is a gradation of climate between the *Sotavento*, which may be considered to have a more typical 'Mediterranean climate' and the *Barlavento*, where the temperatures are lower and there is considerably more wind. The Concelho of Loulé divides the two. The coastal strip running from Cape St. Vincent and the Sagres peninsula north to Odeceixe facing west has a climate rather different from that of the rest of the Algarve: it is windier, colder, rather wet, and much cloudier, with an average of 28 days overcast weather annually.

Within this general classification of Algarve climate, there are major differences between the *Serra*, *Barrocal*, and *Litoral*. The average annual temperature for the major part of the province lies between 15.5°C and 17.5°C. With increasing altitude, this declines, until at 900 m it is 13°C. From April to September inclusive the average temperature increases to 30–35°C, sometimes exceeding 40°C. From January to March, temperatures as low as –6°C have been recorded. Such temperatures place limits on the natural vegetation cover and also on the type of cultivation which can be practised. The danger of frost limits production of subtropical and tropical fruits, posing special difficulties for growers of avocados (*abacates*), for example. Another major factor affecting the crops which can be grown is the wind, so there are few commercial plantations of citrus in the western *Barlavento* for example. The number of hours of sunshine in the *Barrocal* and the *Litoral* is approximately 3000 per annum, of which there are about 1100 hours in the months of October to March inclusive, but in the *Serra* the total is less, some 2500 hours per annum.

Hydrology

The area covered by the rivers in the province is some 3848 km^2. These rivers are the Alcantarilha, Aljezur, Almargem, Arade, Bensafrim, Bordeira, Gilão, Odelouca, Odiáxere, Quarteira, and Seixa, as well as, shared with Spain, the Guadiana.

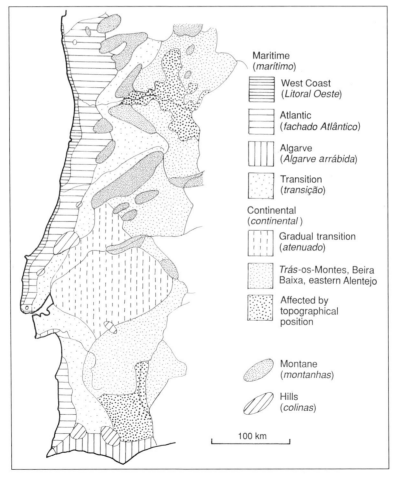

Figure 14 Climatic regions of Portugal. *Mapas climáticos de Portugal.* (Redrawn from Daveau 1985, p. 25.)

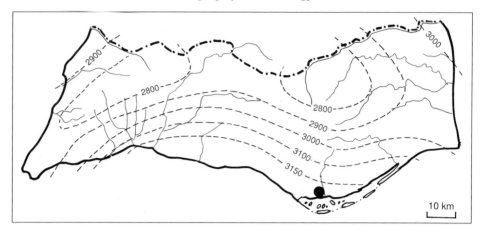

Figure 15 Sunshine (annual number of hours of insolation) (*Sol*). (Redrawn from Rocha Faria *et al.* 1981.)

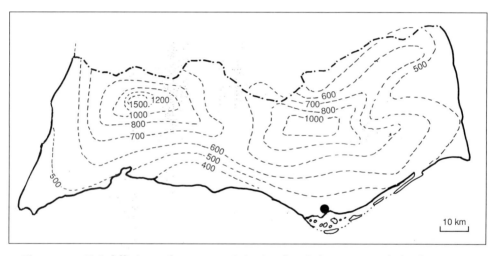

Figure 16 Rainfall: Annual mean precipitation (mm). (*Precipitação*). (Redrawn from Rocha Faria *et al.* 1981.)

Rainfall

The average annual rainfall in the Algarve in the 90-year period 1895–1984 was 466 mm. This disguises major variations, both annually and by geographic zone, with annual rainfall often being 2000 mm in the mountainous areas such as the Serra de Monchique, but less than 50 mm at Vila Real de Santo António in the *Litoral*, for example. Evaporation averages 500 mm per annum in the province, although again there are wide variations, 2500 mm being recorded at Figueira and 41 mm at Vila do Bispo. In general, the *Barrocal* and the *Litoral* have much less rain than the higher *Serra*, and in the summer months, May to September inclusive, virtually no rain at all. The average values conceal very large local variations. For example, between 1968 and 1969, the Monchique area received on average 2046 mm per annum, the Arade

Figure 17   River Arade between
Silves and Portimão.

Figure 18   River Odelouca.

Dam 1039 mm and Tavira, on the coast, 1018 mm. However, in the 'dry' years of
1980 and 1981, the rainfall was only 682 mm in Monchique, 224 mm at the Arade
Dam, and 266 mm at Tavira.

   Since evaporation can exceed 1000 mm per annum, there is a large water deficit in
some areas which, in the case of cultivated crops, can be met only with irrigation. As
long ago as 1811, it was notable that a town like Albufeira (ancient Baltum) relied on
water being brought in from the countryside, a problem compounded by the appalling
road conditions of the time. It follows that the distribution of water, as well as the
absolute amounts required, poses a large-scale problem for the maintenance and development of commercial agriculture in the Algarve. The water deficit can only be met, if
the demands of tourism and agriculture are to be reconciled, by an effective overall
water management policy which allows for a substantial increase in the reservoir
capacity in the *Serra* where the average rainfall is at its highest and evaporation
lowest. Additional strategically placed boreholes can also be of much value.

Reservoirs

Barragems were built in the Algarve by the Romans; remains of significant ones
survive at Álamo (Alcoutim), Fonte Coberta (Lagos), Santa Rita (Vila Real de Santo

António), and Vale Tesnado (Loulé), whose hydrographic basin covers some
37.5 km². Since there are no major freshwater lakes in the province, further reservoirs
are being built and others are at the planning and construction stage. In the eastern part
of the province, the reservoir at Beliche has a capacity of 44 million m³. The Funcho-
Odelouca reservoir near Silves, with a capacity of 43 million m³, nearing completion
in 1992, is the largest in the Algarve. The Barragem da Bravura in the west has a
capacity of 33 million m³ and, in addition to some electric production at the outfall,
there are 20 km of large canals and a further 28 km of aqueducts leading to 68 km of
smaller irrigation waterways as well as associated drainage ditches. This extensive
system irrigates the land between Portimão and Lagos, although many of the
distribution channels have now fallen into disrepair and disuse. In all, irrigation water
is supplied to 1800 ha of land as well as drinking water to nearby towns. Near Alvôr,
some salt-marshes, above mean sea-level but affected by tidal movements, are dyked
and have to be drained by pumps, as sea-water seeps through the porous rocks of the
soils in the *Litoral*.

Generally speaking, salinity is increasing throughout the Algarve, but especially in
the *Litoral*, owing to deep drilling of boreholes coupled with sea-water filtering back
there.

Other major *barragems* include Arade, north of Silves, which supplies irrigation
water to 1900 ha of land. Further large reservoirs are planned for Odeleite, Vascão,
and at Foupana, all three in the eastern Algarve. Their construction will go some way
to satisfying the extra demands for water and might help to counter the increasing
salinity of water drawn from the numerous boreholes in the region. The three new
reservoirs are intended to supply the irrigation needs of some 50 000 ha of land in the
Barrocal and the *Litoral*. These regions have about 60 000 ha of *mato*, the scrub
vegetation typical of the Algarve, with potential for productive agriculture, but which
today is virtually unused. By 1990 about 4000 ha were being irrigated from the

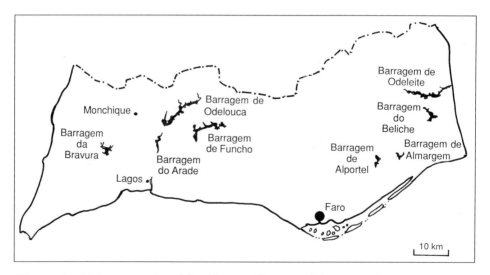

Figure 19   Major reservoirs of the Algarve. *Plano geral dos aproveitamentos
hídricos superficiais do Algarve: grandes sistemas.* (Redrawn from Ramos *et al.*,
1988, Fig.30.)

barragems, with a further 18 000 ha from wells (*poços*) and boreholes (*furos*). For example, 10 000 ha of citrus and 1000 ha of table grapes are irrigated from these combined sources, as are 5400 ha of horticultural enterprises in the open, and a further 800 ha of crops grown in plastic-covered greenhouses. Many new smaller reservoirs are also being built in the folds of the *Serra*, although these will provide water only for very local needs.

Boreholes and wells

For many years there has been inadequate control of borehole drilling. In some areas of the Algarve this has meant that in places where in 1974, for example, water for local distribution could be reached at a depth of 30 m, drilling to 80 m is now necessary to secure an adequate supply; indeed, the *câmaras* (local councils) and others are now drilling down to 200 m. Most of these deep tube wells, with associated stone-clearance, are providing irrigation in the *Barrocal*. Water from such sources is pumped vertically with electric submersible pumps. By contrast, wells are usually 10–15 m in depth. Some of these date from the eighth century, and they involved technology found in many of the lands bordering the Mediterranean whereby water was drawn by mules or donkeys pulling a chain of buckets on a wheel set vertically (*noras*). These wells are now mostly disused, though some have been connected to electric pumps.

Water management, and indeed much detailed planning, is controlled by the *câmaras* and the central co-ordinating water authority has limited resources to enable

Figure 20   Irrigation system near Alcalar.

it to plan, invest, and allocate the limited resources. One consequence is that water supply costs to consumers are among the highest in Europe, and the average purity is low. Another is that since higher priority is given throughout the Algarve to the development of tourism and its needs for ever-larger quantities of water, there is inadequate supply for agricultural expansion in areas of the province which otherwise have not only a suitable climate but also favourable soils. However, the physical problems and costs of distributing water, and electricity for power as well, in adequate volumes and of sufficient quality to large numbers of small farms are considerable, as some 72 per cent of these are smaller than 4 ha.

Figure 21   Chain bucket well. *Nora.*

Springs

Finally, mention should be made of the mineral water springs which are found in the Algarve. The most important, as it has been since Roman times, is that at Caldas de Monchique. The water there is sulphurous and weakly fluorinated according to the classification established by the *Instituto de Hidrologia de Lisboa* and re-confirmed by recent work using modern methods of numerical taxonomy. However, there are many other springs, for example at Atalaia (Tavira), Felgueira, and Gerês. Whilst all these are strongly alkaline, the waters differ significantly in chemical constitution and have been greatly prized for their therapeutic value, as have those at Alte and Alcoutim.

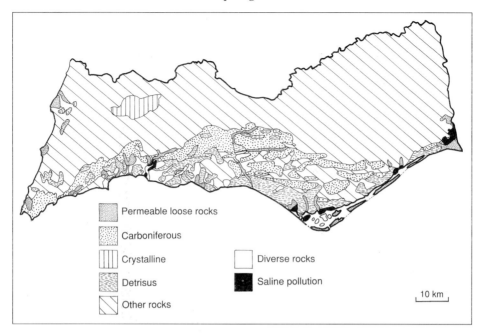

Figure 22   The Algarve: hydrogeology. *Aptidão hidrogeológica.* (Redrawn from Comissão de Coordenação da Região do Algarve 1985, Anexo B 7.2.)

Legend:
- Permeable loose rocks
- Carboniferous
- Crystalline
- Detrisus
- Other rocks
- Diverse rocks
- Saline pollution

10 km

Figure 23   Washing clothes in river near Alvôr.

Geology

The earliest detailed description of the geology of the Algarve is that of Bonnet published in 1850. More recent work by the *Serviços Geológicos de Portugal* (SGP) has greatly improved our knowledge. Broadly, the geology follows the geography in that there are three distinct parts, corresponding to the major geographical features of the *Serra, Barrocal*, and *Litoral*. These are respectively as follows:

(1) in the southern 'strip' of the Algarve, the heterogeneous zone of sedimentary rocks, alluviums, and sands (of the *Litoral*);

(2) the contiguous central limestone zone (of the *Barrocal*);

(3) the relatively homogeneous shale zones extending across the mountainous, *Serra*, northern part of the province; this area includes the syenitic, crystalline rock, intrusion of Monchique.

The geology of the Algarve is rather variable, especially in the *Litoral*. None the less, the following geological eras have been recognized as being those most important in the chief components of the three zones.

Quaternary (Quaternário)

These deposits are up to two million years old, and are a major part of dunes, edges of rivers and the sea, alluvial plains, and so on. Hence such formations are especially associated with the *Litoral*.

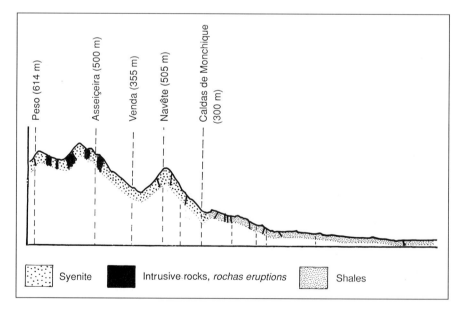

Figure 24   The Monchique massif: transverse section. *Corte transversal da Serra de Monchique.* (Redrawn from Silva Carvalho 1939, p. 184.)

Tertiary *(Terciário)*

Tertiary formations are from two to 60 million years old. Rocks of this age can be seen very clearly by the sea at Ponta da Piedade (Lagos). In the main, they are associated with sedimentary deposits found in coastal regions, namely the *Litoral*, but are also found in areas where there has been major tectonic movement, for example the syenitic intrusion of Monchique.

Cretaceous *(Cretácico)*

These rocks are from 60 to 130 million years old and are also found as sedimentary rocks in coastal areas, as agglomerates, quartzite, dolomitic limestone, and multi-coloured shales. These sediments and the soils associated with them are closely identified with the *Litoral* region, where they can be easily seen to the east of Praia da Luz for example. They are also quite common in the *Barrocal*.

Jurassic *(Jurássico)*

These have an age estimated at 133–135 million years. Typically these rocks are dense, often dolomitic, limestones commonly found in the *Barrocal* and at altitudes of up to nearly 1000 m. However, some rock formations between Cape St. Vincent and Sagres are of this type, for example by the beaches at Sagres itself and at Enseada de Beliche.

Triassic *(Triássico)*

These rocks are typical of regions at an altitude of 100–400 m above sea level, for example in the Silves sandstones. They are also found in fertile valleys and in the lower *Serra*.

Carboniferous *(Carbónico)*

These formations are thought to have been laid down 210–265 million years ago. They are found in the higher parts of the *Serra*. Rock formations to the north of Vila do Bispo, for example around Castelejo on the west coast, are of this type.

Economic geology

In some parts of the Algarve, where stone is available, dry walls, traditionally coated with '*cal viva*', quicklime from pure dolomitic limestone, for field and other boundaries, are made of rock, as are the foundations and walls of many older houses. Until the 1960s, there were very large accessible quantities of soft sandstone on or close to the surface of the ground in the *Litoral*. The large cement factory near Loulé depends on local supplies of gypsum, which is quarried from the limestone formations of the *Barrocal*. Large quantities of harder stones are quarried, for example north-east of Guia, much of this associated with the very great expansion in construction necessitated by the tourist industry.

There are a number of iron-bearing strata in the north-eastern part of the Algarve, centred on Leitejo. Other ironstone seams run through the province, many of them close to the surface. These deposits led to the settling of ironsmiths, many of whom are still to be found around Silves, Portimão, Loulé, and Tavira. Similar ironstone deposits exist in Baixa Alentejo, the southern part of the province to the north. Slate has been mined at Corte Cabreira near Aljezur. Also in this area, Vidigal is the site of ancient copper mines. Lagos, and even more so Loulé, are the traditional places where the arts of the coppersmith and tinsmith have been practised. Many remnants of copperworkings are also to be found in the *freguesia* of Monte de Castelo (Estoi). Besides a wide range of decorative wares for general household use and for the tourist industry, hammered copper has been produced for many centuries to make the *cataplana*. This is an important cooking utensil consisting of two nearly equal and flattened hemispheres which can be clamped together and sealed, so functioning as a pressure cooker, although much simpler in construction and use. The principal dish for which it is used is a mixture of clams and pork cubes, which are cooked together with herbs, *chouriço* (salami), onions, and tomatoes. Whilst copper and tin are no longer mined in the Algarve, Europe's biggest mine for both metals became operational in 1989 in southern Alentejo at Neves Corvo, with a production in 1990 of 660 000 t of copper concentrates and 4700 t of tin concentrates. Neves Corvo, with its reserves

Figure 25 Carboniferous shales at Amoreira.

Figure 26 Stone wall near Penina, *Litoral.*

Figure 27 Neglected stone wall showing structure.

exceeding 33 million t of copper concentrates and 300 000 t of tin concentrates, could lead to a regeneration of the ancient metalworking arts in the Algarve.

Soils

The Algarve lies in a zone which escaped the worst effects of the last major glaciation, which ended about 8500 years ago. At that time, the ice and cold of Northern Europe did not correspond generally with a very wet period in Southern Europe as was once thought, but recent pollen analysis studies indicate a cold and dry period with poor vegetation and a cover of *loess*, wind-blown and fertile dust found in the valleys. Soil types found in the Algarve are much influenced by erosion from the *Serra* where, for many centuries, materials have been carried down from the higher slopes, filling the valleys; in turn, these are washed into the sea. Such movements have been most evident in times of climatic change and have been hastened by poor husbandry, such as ploughing down the slope and by the removal of consolidating roots.

Some 200 different types of soil have been characterized in the province, the diversity resulting from the past climate together with the effects of man and his animals.

Figure 28   Coppersmith at Loulé. (Photograph Travelpress Europe.)

There is considerable variation in the acidity of the soils, even within the western Algarve. Following the geographical and geological features already described, these soils range from very chalky to very acidic. The significance for plant life is that many species are not lime-tolerant and therefore are found only in the acid soils of the *Serra* and parts of the sandy *Litoral*.

Of the various types of soil found in the province, those in the *Serra* are generally very thin, acidic, low in nutrients owing to erosion, and with many loose stones, often on steep slopes. Consequently, even after terracing, utilization of these mountainous areas in the past has been largely restricted to forestry, recently particularly to eucalyptus, but with substantial quantities of pine and cork oak, and some chestnut. The forest areas of the *Serra* make up about 62 000 ha, of which 26 000 ha are given over to cork oak, 14 000 ha to carob, 11 000 ha to eucalyptus, 4500 ha to pine, and 4000 ha to the oak, *Quercus ilex* ssp. *ballota* (*Q. rotundifolia, azinheira*). The areas intensively cultivated in the *Serra* for agriculture, including forestry, total about 30 000 ha.

Cereals are also grown in the wetter zones, where a programme of planting almonds and carob has been initiated. The carob is mainly associated with the production in the same area (from strawberry trees, see Chapter 4) of *medronho*, a distilled spirit. With the construction of many small water reservoirs, faced with earth and 'puddled-in' with clayey soils to ensure water retention at minimum construction cost, fruit trees have been planted in many of the valleys. Flocks of goats browse and sheep graze extensively in the *Serra*, whilst bee-keeping is also important there. By contrast, the limy soils of the *Barrocal* support in the main the traditional trees of the Algarve, that

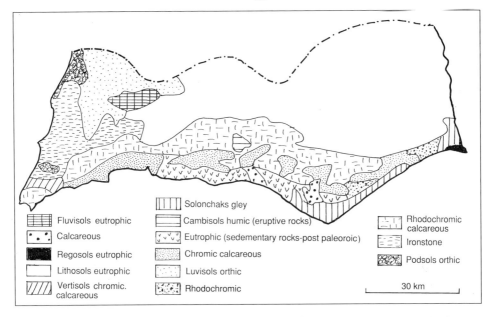

Figure 29 Soil types in the Algarve. *Os solos do Algarve e as suas caracteristícas.* (Redrawn from Rocha Faria *et al.* 1981.)

is to say fig, almond, olive, and carob. With all these trees, there are various crops grown under the cover of the tree canopy and in between plantings. In the *Litoral*, grapes, for both winemaking (*uvas para vinho*) and for the table (*uvas de mesa*), are very commonly grown, as is citrus, notably around Silves. Horticulture and floriculture are important industries, often using *estufas* (plastic-covered greenhouses).

There are about 104 000 ha of agricultural land in the *Barrocal* and the *Litoral* and of this area, about 94 000 ha are planted with fruit trees, of which 70 000 ha (including 3000 ha of vines) are the traditional almond–carob–fig–olive complex and 22 000 ha are irrigated, mainly for citrus but with some peaches, nectarines, pears etc. There are some 5 600 000 almond trees and 950 000 olive trees. About 14 000 ha are interplanted with cereals, and a further 3000 ha with crops such as peas and broad beans. Horticulture, much of it centred on Faro, occupies about 10 000 ha.

Ecology

Although the forest cover of the country is much reduced, 3 234 000 ha or 36 per cent of the land surface overall, the vegetation of Portugal may be classified according to the dominance of particular oak species. In the north, there is the deciduous English oak (*carvalho alvarinho, Quercus robur*, Fagaceae). The southern part is notable for the predominance of evergreen oaks, typical of the Mediterranean. In 'Mediterranean' Portugal as a whole, 16.6 per cent of the land surface is forest (of which 7.7 per cent is coniferous and 0.9 per cent burnt areas), 39.2 per cent is shrubland, 31.8 per cent is under crops, and 12.4 per cent pasture.

Figure 30   Carobs with grazing animals.

Although not in the Mediterranean in a cartographic sense, the vegetation of southern Portugal is typically 'Mediterranean' in character. Such vegetation differs from that elsewhere in Europe in being characterized by evergreen trees and shrubs which are adapted to survival in the long and rainless summer. Herbaceous plants are either ephemeral, withering with the onset of the summer heat or, if perennial, dying back to underground buds like bulbs or corms. The adverse season is the summer and the main growing period begins with the autumn rains; for example, carobs flower and the sea-squill (*cebola albarrá, Drimia (Urginea) maritima*, Liliaceae), puts up its elegant spikes of flowers then and, by the end of November, the beautiful *Iris planifolia* and the paperwhite narcissus *Narcissus papyraceus* are coming in to bloom.

The distribution of the Mediterranean vegetation more or less coincides with that of the olive (*oliveira, Olea europaea*, Oleaceae), which was spread to Portugal by humans. The effect of human beings, and of their animals (see Chapter 8), in the Mediterranean region has led to the rural landscape being dominated by evergreen shrublands tolerant of fire and grazing. These shrublands are known as *maquis* (*matos* in Portugal). None the less, semi-natural forest vegetation characterized by the presence of pines and other evergreens, particularly oaks, persists when they are of continued use to humans. Under increasing pressure, the high *matos*, which may attain 3–4 m or more in height, are replaced by lower shrublands or even by exposed rock with spring-flowering scattered herbs. Under sustained farm management (see Chapters 4 and 5) the shrubby vegetation is replaced by arable or pasture lands. In southern Portugal, traditional tree crops make up an agroforestry system which is typical of much of the rural landscape of the Algarve.

The Portuguese flora

There are about 2500 species of vascular plants — flowering plants, including conifers and their allies, and ferns and their allies — found growing wild in Portugal. Not all of them are native, many very conspicuous plants typical of the landscape being planted (woody and herbaceous crops), or escaped exotics like the 'Bermuda buttercup' (*erva pata*, *Oxalis pes-caprae*, Oxalidaceae). This plant is native in southern Africa, but is ubiquitous in much of the Mediterranean and elsewhere. According to the latest estimates (1991) of native plants, there are five species of vascular plant (as well as one subspecies of *Ulex argenteus* (Leguminosae)), restricted to the Algarve. Of these, *Tuberaria major* (Cistaceae) is known only from the Algoz and Faro areas, and the rest are found in the south-west coastal region. A campion, *Silene rothmaleri*, collected only once and possibly extinct but perhaps allied to *S. hifacencis* (Caryophyllaceae), and the bluebell *Hyacinthoides vicentina* (Liliaceae), considered by some to be conspecific with *H. italica* of south-eastern France and north-west Italy, are both restricted to the area of Cape St. Vincent, and a toadflax, *Linaria algarviana* (Scrophulariaceae), a blue-and-yellow-, or blue-and-white-flowered species, is a tiny plant of dry coastal sites, with *Sideritis algarviana* (Labiatae) on the *Barrocal* near Lagos.

A further ten species and four subspecies or varieties of plants restricted to Portugal are also found in the Algarve, again mostly in the coastal areas. These plants include *Biscutella vincentina* and *Ionopsidium acaule* (both Cruciferae), a yellow-flowered toadflax *Linaria ficalhoana* (*tomilho-de-cabacinha*), *Thymus camphoratus*, and *T. lotocephalos* ('*T. cephalotos*'), the last-named often found with *Tuberaria major*. Besides these, there are almost another 50 species and subspecies native in Portugal but restricted in their distribution to the Algarve. These include many plants also to be found in North Africa. A prominent example is the dwarf palm *Chamaerops humilis*. Others include the prickly *Astragalus tragacantha* ssp. *vicentinus* (Leguminosae), the parasitic *Cynomorium coccineum* (Cynomoriaceae), *Limoniastrum monopetalum* (Plumbaginaceae) of the salt-marshes, the esparto grass *Stipa tenacissima*, and *Viola arborescens* (Violaceae). Almost 50 more of the plant species are restricted to the Iberian peninsula. These include *Paeonia broteroi*, named after the first professional botanist resident in the country, Felix d'Avelar Brotero (1744–1828) and *Fritillaria lusitanica* (Liliaceae). A further 25 species, for example *Ranunculus gramineus*, are restricted to Europe.

Early phytogeographical work on the flora of Portugal was published by Heinrich Moritz Willkomm (1821–95), who visited the Algarve in 1844. This has been subsequently modified such that a number of 'elements', here illustrated with Algarve plants, can be recognized.

1. Central and western European

Cytisus (Sarothamnus) scoparius (broom, Leguminosae), *Calluna vulgaris* (white heather, Ericaceae).

Figure 31 *Linaria algarviana*, one of the few Algarve endemics.

Figure 32 *Viola arborescens.*

2. Mediterranean

Quercus suber (cork oak) and *Quercus coccifera* (Fagaceae), *Erica arborea* (Ericaceae), *Rosmarinus officinalis* (rosemary, *alecrim*, Labiatae), *Phillyrea angustifolia* (*lentisco-bastardo*, Oleaceae), *Daphne gnidium* (*trovisco-fêmea*, Thymelaeaceae), *Pistacia lentiscus* (*aroeira*, Anacardiaceae), *Nerium oleander* (oleander, *loendro*, Apocynaceae), Cistaceae. Also here is *Arbutus unedo* (strawberry tree, *medronheira*, Ericaceae), which extends to the British Isles where it is often considered part of a 'Lusitanian Element'. This shows that the concept of such elements often reflects the starting-point of the writer.

3. Macaronesian (Atlantic Islands)

Corema album (Empetraceae).

4. Iberian-Atlas plants

Drosophyllum lusitanicum (Droseraceae), *Stipa tenacissima* (Gramineae).

5. Those restricted to Iberia

See above.

6. Those restricted to Portugal

See above.

The native flora can also be analysed according to the life form, there being the geophytic plants mentioned above, as well as succulents such as *Sedum* spp.

Figure 33 *Ranunculus gramineus.*
(Photograph P. Garton.)

(Crassulaceae). Additionally, there are the 'switch' plants which have photosynthetic stems and reduced leaves, such as *Retama monosperma*, broom, and *Genista (Chamaespartium) tridentata* (Leguminosae). These are especially suited to the tough conditions of the Algarve summer. Many are highly armed legumes such as *Astragalus tragacantha* ssp. *vicentinus* and *Ulex* spp. as well as *Asparagus albus* (Liliaceae), or are very aromatic and disagreeable such as *Thymus* spp., *Bituminaria bituminosa* (Leguminosae), and *Ruta* spp. (Rutaceae), which are deterrent to many grazers.

There are a number of woody climbers or lianes, some armed like *Smilax aspera* (Liliaceae) and some herbaceous like *Tamus communis* (*uva de cão*, Dioscoreaceae). There are parasites such as species of *Orobanche, Cistanche phelypaea* of the salt-marshes (Orobanchaceae), and *Cytinus hypocistis* (*pútega*, Rafflesiaceae) of the *matos*. There are carnivorous plants including *Utricularia australis* (Lentibulariaceae) of wet areas, and the woody *Drosophyllum lusitanicum* of pine woodlands and other dry areas. 'Lower plants' are less conspicuous in the flora but specialists have worked on these, and some of their publications are listed in the bibliography.

Algarve vegetation

Vegetation can be classified according to the predominant life forms — forests, shrublands etc., or according to overriding features such as physical factors, e.g. beach vegetation, riparian vegetation etc., or according to the assemblages of plant species themselves. Many classifications are in effect amalgams of such analyses. That of the Algarve has been subdivided largely according to the principal geographical and

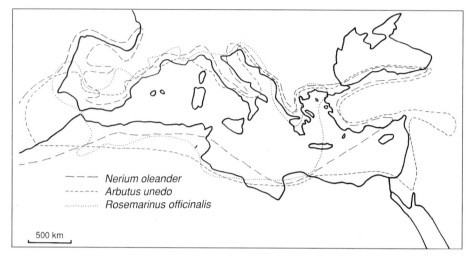

Figure 34   Distribution of some 'Mediterranean' plants. *Distribuição de algumas espécies vegetais do Mediterrâneo.* (Redrawn from Lautensach 1988, p. 543.)

geological features corresponding mainly to distinctive plant assemblages, although these are by no means exclusive such that under the prevailing human regime, species such as *Lithodora prostrata* can be found almost throughout. These are as follows:

(1) *Litoral* (south)

Figure 35   *Tamus communis.*

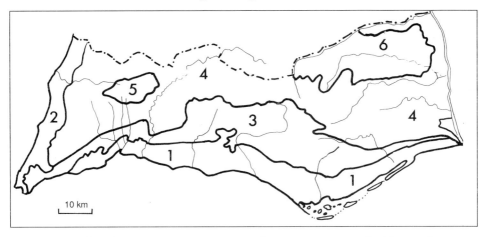

Figure 36 The Algarve: vegetation types. *A flora e a vegetação do Algarve*. See text for explanation. (Redrawn from Dias 1986, p. 706.)

(2) *Litoral* (west)

(3) *Barrocal*

(4) *Serras do Algarve*

(5) Monchique

(6) *Planalto.*

In an Iberian context, the first three are considered part of an Algarvian floral province extending into southern Spain as well as into coastal Alentejo. The last three are part of an upland province extending far into Spain as well as into much of Alentejo.

Against this background, the plant life of rural Algarve will be dealt with from maritime to montane, and, in an ecological context, land use and rural industries will be considered. This is followed by a brief account of urban plants and landscapes. In this way, we present an account of the plants and landscape of the Algarve as visitors find them. Therefore, we stress that which is most apparent and catches the eye, rather than provide an exhaustive survey of the botany of the region.

2

The coastline

Sea-fishing

There is a long tradition of sea-fishing, as fish in the Algarve is a readily available source of fresh high-quality protein. The demand is increased by the Roman Catholic practice of 'fish-only days' followed by much of the population. Famous Algarvian fish dishes include fish-stew (*caldeirada*), a mixture of small pieces of cod, diced potatoes, and onions known as *bacalhau à Bráz*, grilled tuna (*atum grelhado*), and smoked swordfish (*espadarte fumado*).

The major fishes caught off the Portuguese coast are listed below:

Sardinha (sardine, *Sardina pilchardus*)   46.0 per cent
Carapau (horse mackerel, *Trachurus trachurus* (*pequeno*)
and *T. picturatus*)   6.0 per cent

Figure 37   The harbour at Sagres.

Figure 38   Fishing boats at Ferragudo.

Peixe-espada (scabbard fish, *Lepidopus caudatus*)	5.5 per cent
Pescada (hake, *Merluccius merluccius*)	5.5 per cent
Cavala (mackerel, *Pneumatophorus japonicus*)	4.0 per cent
Faneca (whiting (pout), *Trisopterus luscus*)	3.0 per cent
Chicharro (large horse mackerel, *Trachurus trachurus* (*grande*)	3.0 per cent
Safio (conger eel, *Conger conger*)	2.0 per cent
Raia (skate, *Raja* spp.)	2.0 per cent
Pargo (sea bream, *Pagrus pagrus*)	2.0 per cent

Total 79.0 per cent (by weight)

with about a further 100 species making up the total, of which the bulk are *atum* (tunnyfish, the three species caught off the Algarve being *Orcynnus thynnus*, *O. alalonga* (*albacora*, albacore), and *O. albacora* (*atum albacora*)), *besugo* (sea bream, *Pageullus acarne*), *boga* (bogue, *Boops boops*), *corvina* (corvina (weakfish), *Pseudotolithus* sp.), *dourada* (John Dory, *Sparus aurata*), *enguia* (eel, *Anguilla anguilla*), *espadarte* (swordfish, *Xiphias gladius*), *linguado* (sole, *Solea lascaris*), *pregado* (turbot, *Scapthalmus maximus*), *robalo* (sea-bass, *Dicentrarchus labrax*), *salmonete* (red mullet, *Mullus surmuletus*), sargo (*Diplodus sargus*), *solha* (plaice, *Placichthys flesus*), and *tamboril* (monkfish, *Lophius piscatorius*).

The catch also includes the very unusual scabbard fish (*peixe espada-branco*) and sometimes its Madeiran counterpart, the *peixe espada-preto*, which are caught on long lines at depths greater than 400 m, the pressure difference killing them as they are hauled to the surface. The scabbard fish is quite different from the *espada* (swordfish) of Spain, the Portuguese swordfish being *espadarte*.

Figure 39 *Pargo.* (Photograph Travelpress Europe.)

Figure 40 Landing sardines at Portimão.

Figure 41   Drying eels at Ferragudo.

The major fishing ports are Olhão (19 000 t per annum) and Portimão (17 000 t per annum), minor ports being Vila Real de Santo António (about 4400t (1986)), about the same as in 1978, although it was smaller in the intervening years. The Algarve catch now represents about 20 per cent of the total (about 210 000 t) caught off Continental Portugal. Most is landed from vessels less than 14.5 m long, trawling within 5 km of the coast. There is some deep-sea fishing with larger boats from Olhão, the practice dating from the fifteenth century when a treaty with Spain gave Portugal the exclusive rights on the African coast between Cabo Bojador and the Rio do Ouro. Long-distance fishing for cod (*bacalhau, Gadus morrhua*), initially from the port of Aveiro on the west coast of Portugal in the sixteenth century, took the Portuguese to the grounds off Newfoundland.

In the eighteenth century the enlightened policies of the Marquês de Pombal, who established Vila Real, led to a resurgence in fishing in the Algarve, firstly of sardines and then later of tunny. Since then the pattern of fishing has been much affected by several factors.

1. In 1831 the monopolies for tunny, a very large and valuable fish up to 350 kg in weight and 2 m long at 20 years old, and *corvina* were liberalized. Since 1249, tunny-fishing had been a Crown monopoly, and a provincial one in the Algarve from 1773 onwards, and tunny were known as *Pescarias Reais* (Royal Fish). At first this led to increasing catches, but later to over-fishing and a decline followed by a reduction in the availability of tunny and, in turn, by both the disappearance of salting and a significant reduction in canning. The 42 000 tunny landed in 1914 were reduced to an average of fewer than 1000 per annum by the 1960s.

2. In the 1930s there was a decline in the availability of sardines, nearly half by weight of the annual catch of all fishes in the Algarve, and this in turn caused the disappearance of salting and a decline in canning.

3. There have been large variations in exports. Sales of fish to Spain in particular were very important from medieval times until early in this century; for example, the recorded total of sardines exported to Spain in 1870 was 5400 t. Moreover, there can be little doubt that fishing boats from the Guadiana were able at various times to evade bothersome controls in Portugal by clandestinely transferring their cargoes either at sea or at Ayamonte on the Spanish side of the river.

4. In this century fishing has been severely affected, not only by the two World Wars, but also by the Spanish Civil War with its aftermath and the economic crisis of the 1930s. Moreover, since the 1950s the fishing industry has undergone much restructuring, led by improvements in techniques such as the widespread use of trawlers, known as *traineras*, specially equipped for catching sardines.

5. Recently, European Community fishing quotas have placed further limits on the expansion of the Portuguese fishing industry.

Molluscs and other invertebrates

Shellfish are plentiful in the Algarve, especially
Amêijoas (clams, *Ruditopes decussatus*),
Berbigão (cockles, *Ceratoderma edule*),
Camarão (shrimps, *Aristeus antennatus*, *Crangon* spp., *Palaemon* spp.),
Chocos (cuttlefish, *Sepia officinalis*),
Conquilhas (*Donax trunculus*),
Lagosta (rock-lobsters, *Palinurus mauritanicus*),
Lagostim (Norwegian lobsters, *Nephrops norvegicus*),
Lavagante (American lobsters, *Homarus gammarus*),
Lulas (squid, *Loligo vulgaris*),
Perceves (barnacles, *Pollicipes cornucopia*),
Polvo (octopus, *Octopus vulgaris*), and
Santola (spider crabs, *Maja* spp.).

Although sea-fishing generally is at best economically stable, the quantity of high-value crustaceans caught is increasing. Molluscs harvested in 1989 in the Algarve included 10 000 t of *berbigão* and 4000 t of *amêijoas*, much of this around Faro and at Alvôr, though there production is declining in part owing to the effects of dredging and pollution.

A positive trend in the 1980s in the Algarve has been the development of aquaculture, especially in Ria Formosa, with the annual yield from about 2100 ha of ponds now exceeding 8000 t, which is 80 per cent of Portugal's production. Because of disease, the culture of the Portuguese native oyster, *ostra*, *Ostrea edulis*, which formed the basis of Europe's largest oyster production until 1950, has now almost ceased. Most raised in the Algarve today are clams, though in 1985 many were destroyed by protozoan parasites.

Figure 42   Rocky shore at Amoreira, mussels habitat.

Figure 43   Harvesting mussels at Amoreira.

Figure 44   Collecting cockles at Alvôr.

Figure 45   Collecting cockles at Alvôr.

Salt-marshes

The great salt-marsh area of the Algarve is in the Faro region, although there are smaller areas elsewhere in the estuaries of the Rio Arade and the confluence of the two rivers Rio de Alvôr and Ribeira do Odiáxere, at Alvôr, and the much more extensive system in the Rio Guadiana. The last is now part of the Castro Marim Nature Reserve (*Reserva natural: Sapal de Castro Marim e Vila Real de Santo António*), whilst the area south of Faro is mostly in the Ria Formosa Natural Park (*Parque e Reserva Natural da Ria Formosa*).

Figure 46   Salt-marsh, Ria Formosa.

In the shallows is the sea-grass *Zostera noltii* (Zosteraceae). Starting from the sea, the belts of vegetation across the salt-marsh are firstly the cord grass *Spartina maritima* (*morraça*) (as it is around all of Portugal's coastline), then the glasswort, *Sarcocornia* (*Arthrocnemon*) *perennis* (Chenopodiaceae). This is followed by one of *S. fruticosa* (*gramata*) and *Atriplex portulacoides* (*gramata branca*) of the same family. In the Guadiana, these last two are replaced by *Spartina densiflora* introduced from South America. In the north of Portugal there then follows a zone of the rush *Juncus maritimus* (Juncaceae), but in the south of the country belts of *Arthrocnemon macrostachyum* (*A. glaucum*, *sapeira*) and/or *Suaeda vera* (*valverde dos sapais*) are interpolated. Sometimes this sequence includes *Limoniastrum monopetalum* (*marisma negra*), not otherwise encountered around Portugal's coasts. It often grows by artificial dykes or at the edges of salt-marshes where there is a covering of sand from nearby dunes.

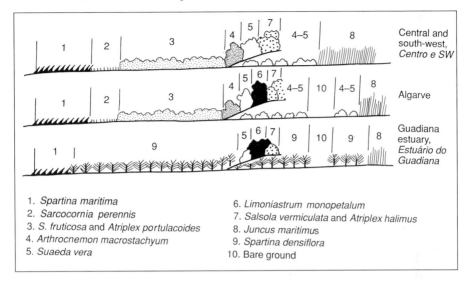

Figure 47 Salt-marsh profiles in southern Portugal. *Perfis de vegetação de sapal.* (Redrawn from Anon. 1988*b*.)

Conspicuous in March and April are the bright yellow inflorescences of *Cistanche phelypaea* (Plate 1) which is parasitic on the roots of the woody Chenopodiaceae, but more rarely seen is the club-like *Cynomorium coccineum*, another parasite. Amongst other plants, for example at Ria Formosa and Alvôr, are species of sea lavender *Limonium* (Plumbaginaceae), notably the feathery-leaved *L. ferulaceum*, and the plantain *Plantago coronopus*. At Castro Marim, the sedge *Scirpus maritimus* is common, as is the prickly dock relation *Emex spinosa* (Polygonaceae). *Arthrocnemon macrostachyum* plants can be invaded by the grasses *Parapholis incurva* and *Hordeum marinum*, later by *Limonium* spp. and *Frankenia laevis* (Frankeniaceae), and finally by the shrubby *Suaeda vera* and *Salsola vermiculata*, with the herbs *Aster tripolium* (Compositae), *Spergularia media* (Caryophyllaceae), *Inula crithmoides* (*madronheira bastarda*, Compositae), and a species of thrift, *Armeria pungens* (Plumbaginaceae).

Salt-pans (*marinhas, salinas*)

Salinas, formerly known as *marinhas*, are salt-pans, traditionally built from packed earth and filled from the sea, the water being evaporated off mainly in the summer months. At high tide, sea-water is allowed to run into a series of shallow ponds separated by narrow dykes. These ponds are separated by walls up to 2 m high, and are often subdivided into smaller drying areas. The liquor is concentrated by the sun and some deposition takes place when the specific gravity of the brine reaches 1.23, but

Figure 48   Salt-marsh plants in the Algarve. *Coberto vegetal tipo em sapais.*
1. *Emex spinosa*; 2. *Limonium ferulaceum*; 3. *Cynomorium coccineum*; 4. *Inula crithmoides*; 5. *Triglochin bulbosa*; 6. *Cakile maritima*; 7. *Spartina maritima*; 8. *Suaeda maritima*; 9. *Cistanche phelypaea*; 10. *Aster tripolium*; 11. *Limoniastrum monopetalum*.

Figure 49   *Salinas.* (Photograph Travelpress Europe.)

crystallization continues until the specific gravity approaches 1.26. When produced in this way, large salt crystals, known as *sal grosso*, are formed. The salt, with mineral impurities contained in it, is then raked into long ridges and allowed to drain, after which it is collected into larger piles, where it is again drained and finally dried. Finer grades of salt are produced from *sal grosso* by mechanical crushing or recrystallization.

Sea-salt (*sal marinho*) production

Production of salt in pans adjoining the sea is a very ancient practice, not confined to Portugal, with a famous example being around the Dead Sea. The Roman army was paid partly in salt, testifying not only to its worth but also to its utility. Such a payment was known as a *salarium*, hence the English word 'salary'. Sea-water contains about three per cent by weight of salt, so that 18 l of sea-water yields about 0.5 kg of salt. Typical annual production from a *salina* of 1 ha is 700 t. Local factors such as the direction of the prevailing winds, the piling up of sand, and the amount and intensity of the sun, affect the production rate, although the orientation of the compartments of the *salinas* is arranged to take into account the first two of these. Algae (*algas*) in the broad sense, notably the 'blue–green' *Microcoleus chthonoplastes* and *Oscillatoria laetevirens*, paving the beds of the *salinas* have a significant effect on the quantity and quality of the salt produced.

Salinas are found along the Algarve coast, and are very prominent at Castro Marim, Tavira, and Faro. The earliest extant Portuguese document referring to salt — and the

Figure 50   Traditional tools used in sea-salt production. *Instrumentos empregados nas marinhas.* (Redrawn from Lepierre 1935, p. 335.)

salinas at Aveiro — dates from 929, whilst in the province there is evidence that this method of working goes back to at least 1314, the year in which Dom Diniz granted the people of Alvôr privileges for the sale of salt: even then, some of these salt-pans were referred to as *marinhas velhas* (old salt-pans). Some authors have ascribed their origin to the Roman period when pans were known as *salgadeiras*. In Lagos, salt was extensively used by 1485 in the export of large quantities of salted tunnyfish (*atum salgado, toninna*), whilst other *salinas* had been established in Faro by 1496 and, by 1532, Dom João III had built 28 *salinas* in Tavira. 'New installations' (of salt-pans) were recorded in 1697 at Castro Marim and, in 1720, the Infante Dom Francisco built *salinas* in Portimão and Alvôr. In 1790, the five *salinas* of Portimão, including Alvôr, were yielding about 2680 *moios* (2010 t) a year and the 185 pans at Castro Marim were said to produce 6240 *moios* (4680 t). At that time the production of sea-salt in Portugal was 286 000 t, most of it near the ports of Lisbon and Setúbal but some also at Figueira da Foz and Aveiro. By 1934 the national total was 210000 t, of which the Algarve accounted for 10 per cent (6000 t each at Faro and Castro Marim, 5000 t at Olhão, 3000 t at Tavira, and 1000 t between Lagos, Portimão, and Lagôa) from a total of 227 *salinas*. By then those formerly at Ferragudo and Mexilhoeira de Carragacão had fallen into disuse. The area (1990) of *salinas* in the Algarve totals 1750 ha, yielding (1988) some 126 000 t per annum.

The production and marketing of sea-salt has been regulated by the Portuguese government since a law of 1576 gave the Crown a monopoly for its sale. Salt is also mined, some 50000 t of rock-salt (*sal-gema*) being extracted annually from a major deposit near Loulé. Salt has been exported from the Algarve for at least two centuries, but most is used locally or in the chemical industry. It has been intimately connected for hundreds of years with the preservation of fish, olives, and pork, all local products.

Dunes

Away from salt-marshes, the coast is often sandy with beaches and sometimes dunes, although, as in the *Barlavento*, the rocks may rise straight from the sea. At Cape St. Vincent, the vertical cliff faces are as high as 80 m. Generally, the sea is very deep at such places. Some 20 km off the south-west coast of the Cape, the water is 1000 m deep, and it rapidly increases to 4000 m. Such depths provide not only a most important ship navigation channel for supertankers, and suitable waters for diving trials of submarines, but are also the habitat of large numbers of fish of many species of great economic importance.

Figure 51   Sand-dune system at the mouth of the Aljezur river.

The beaches are notable for plants with tolerance of shifting sands, strong winds, and variable salinity dependent on precipitation. Moving inland from the sea at Praia da Amoreira (on the Atlantic coast 35 km to the north of Cape St. Vincent) or at Bordeira, the following conspicuous plants are encountered in approximately this order: on the unstabilized dunes are the long rhizomes of the marram grass *Ammophila arenaria* — a species much used in coastal reclamation, putting up tufts of leaves. Then comes the grey–green densely leafy shoots of *Euphorbia paralias* (Plate 2), with the milky-white latex typical of its genus, and with its shoots rising from deeply-buried rhizomes, and the stiff tufts of the grass *Elytrygia juncea*. Following are sea-rocket, *Cakile maritima* (Cruciferae), the prickly leaves of sea-holly, *Eryngium maritimum* (Umbelliferae, Plate 3), the sedge *Cyperus capitatus* (Plate 4), the creeping *Polygonum maritimum*, the succulent-leaved samphire *Crithmum maritimum* (Umbelliferae), and the grey leaves of the sea-daffodil *Pancratium maritimum* (Liliaceae), with white trumpet flowers in July onwards from deeply buried bulbs, depicted in the Minoan frescoes of the Bronze Age (*c.* 1560 BC), and the greyish straggling shoots of *Crucianella maritima* (Rubiaceae), its leaves in four ranks retained as silvery paper husks around the older stems. On the temporarily stabilized gaps in spring are minute prostrate plants of *Corrigiola litoralis* (Caryophyllaceae). In the hollows and ridges of more established dunes, there are the grey narrow-leaved bushes of *Helichrysum picardii* (Compositae), the silvery *Paronychia argentea* (Plate 5), *Anchusa calcarea* (Boraginaceae, Plate 6), *Rumex bucephalophorus* (Plate 7), and *R. intermedius* (Plate 8). In exposed places is the mat-forming succulent introduced from southern Africa *Carpobrotus edulis* (Aizoaceae), with its large dirty-yellow or reddish-pink flowers, the grey-leaved creeping legume *Medicago marina* (Plate 9), with egg-yellow flowers and nitrogen-fixing nodules, the shiny rosettes of rounded leaves of *Calystegia soldanella* (Convolvulaceae, Plate 10) producing pink-and-white-striped trumpets from its brittle rhizomes late in April and May. Other nitrogen-fixing legumes include the green-leaved *Lotus creticus* (Plate 11) and the glistening bright green shrubs of *Ononis natrix* (Plate 12) with large yellow flowers. Between the bushes are the short-lived annuals *Silene littorea* (Caryophyllaceae, Plate 13) with its pink flowers on plants often only 1 cm tall, the daisy *Bellis annua* (Plate 14), and the yellow-flowered *Senecio gallicus* (Compositae, Plate 15), the rosettes of *Centaurea sphaerocephala* (Compositae, Plate 16), and the grey rosettes of *Malcolmia littorea* (Cruciferae, Plate 17), producing large pink–white flowers. As the vegetation passes into low shrubland typical of coastal vegetation on sand, there are species of thrift, *Armeria*, the fescue-like *Corynephorus canescens*, and the shrubby figwort *Scrophularia frutescens*, its flowers visited by wasps in spring, the succulent *Sedum nicaeense*, and species of broomrape (*Orobanche*).

Sand-dune plants have adaptations to high temperatures as well as salinity variation, in that the 'stress metabolites' (proline and glycine betaine) and the polyols (mannitol and sorbitol) which accumulate in plant tissues in response to various climatic stresses ranging from salt exposure to freezing and drought, are also the agents contributing to the heat tolerance of certain enzymes. In *Ammophila arenaria* for instance, the stability of glutamine synthetase increases. Plants under water stress usually have inhibition of protein synthesis and, as the leaves heat up after stomatal closure, there will be a shift in the balance between enzyme synthesis and thermal activation, so that any small increase in enzyme heat stability caused by the 'stress metabolites' leads to a shorter lag before the resumption of full metabolic activity on return to favourable conditions. Plants such as *Carpobrotus edulis* (Plates 18–20) change the nature of their metabolism from that typical of temperate regions in the mild Mediterranean winter to a more drought-resistant system as the summer advances. This change can be induced experimentally by water stress, salt stress, and by change of photoperiod and of thermoperiod. The drought-tolerant system involves stomatal opening and carbon dioxide uptake at night with closure during the day, thereby reducing water-losses, but the tissue temperatures by day may increase to 10–15 °C above the air temperature.

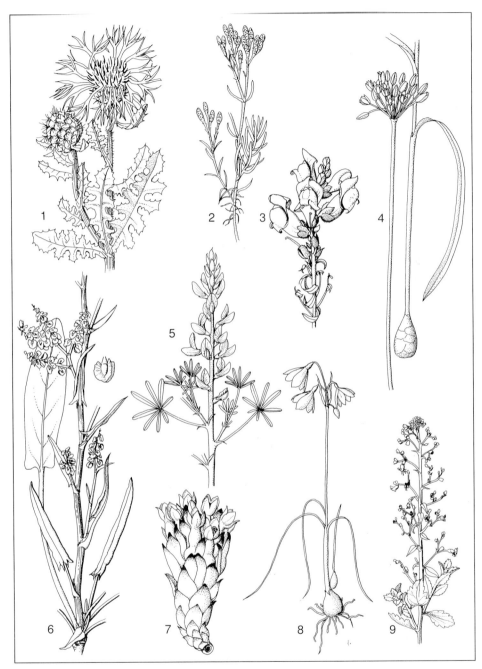

Figure 52   Algarve dune plants I. *Vegetal tipo em estrutares dunares* I.   1.
Centaurea sphaerocephala; 2. *Helichrysum picardii;* 3. *Antirrhinum majus* ssp.
cirrhigerum; 4. *Allium subvillosum;* 5. *Lupinus angustifolius;* 6. *Rumex intermedius;*
7. *Cytinus hypocistis;* 8. *Leucojum trichophyllum;* 9. *Scrophularia frutescens.*

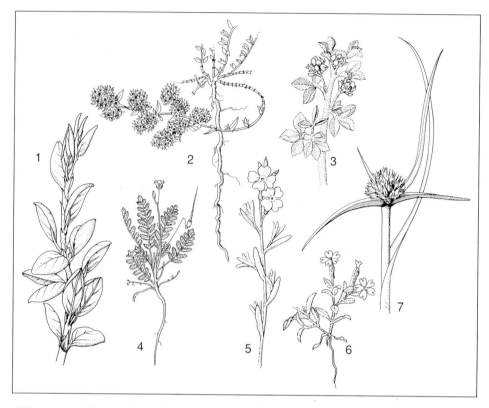

Figure 53   Algarve dune plants II. *Vegetal tipo em estrutares dunares* II. 1. *Polygonum maritimum*; 2. *Paronychia argentea*; 3. *Medicago marina*; 4. *Erodium aethiopicum*; 5. *Malcolmia littorea*; 6. *Silene littorea*; 7. *Cyperus capitatus*.

Succulents such as *Carpobrotus* have a water-storage system reducing the risk of tissue dehydration and excessive salt concentrations. So effective are these mechanisms that the plant is an efficient competitor and is now a danger to the native flora even in south-west Britain as well as in Continental Europe. It was introduced to Britain *c.* 1690 but its date of introduction to the Algarve is unknown. It is not clear whether it was introduced accidentally or for its fruit or other qualities. The fruit, brown and wrinkled when ripe, is edible with a sweet but acidic taste, and can be eaten raw, dried, or in jam. The juice from its leaves has been used for the treatment of dysentery, its antiseptic properties leading to its being used in the treatment of sore throats and, traditionally in southern Africa, it is smeared over new-born babies.

   C. edulis reproduces vegetatively and from seed, fragments collected by gulls for nesting material become established so that it has spread even to uninhabited islands. The mats often overwhelm native vegetation and the litter under the dead and dying stems seems inimicable to the establishment of other plant species: in the main it comprises slowly decomposing dried leaves. *C. edulis* thrives where the vegetation-cover has been damaged and where nutrient levels are apparently high, but it is capable of establishment in the sandy areas between clumps of vegetation at Cape St. Vincent, Amoreira, and elsewhere, where frequently it invades and engulfs other plants. Fires which severely affect cliff-top vegetation permit it to spread further. Its competitive capacity is also promoted by fire where it has been planted in California. In Portugal it is particularly abundant on the Sagres Peninsula, along much of the coast from there up to Sines, and further north at another tourist site, Cabo da Roca, west of Lisbon. Other escaped exotic Aizoaceae in the Algarve include a species of *Drosanthemum* (Plate 21), a shrub from southern Africa now well established in the *Fortaleza* at Sagres and elsewhere in the *Barlavento*, where it produces its lilac flowers.

The dunes at Ria Formosa include the creeping *Polygonum maritimum*, which is conspicuous on other beaches like that at Ferragudo (near Portimão), and many typical salt-marsh plants as well as the familiar dune-plants from elsewhere: *Malcolmia littorea*, *Medicago littoralis*, *Eryngium maritimum*, *Armeria pungens* (Plate 22), *Crucianella maritima*, *Calystegia soldanella*, *Helichrysum picardii*, *Ammophila arenaria*, and *Pancratium maritimum*. Besides the often aggressive *Carpobrotus*, here and in many places around the coast is another southern African exotic naturalized in the dunes, *Arctotheca calendula* (Compositae, Plate 23). The dune systems at the eastern extremity of the Algarve, at Vila Real de Santo António, are similar, but have stands of *Retama monosperma* (Plate 24) in the sheltered hollows, forming an understorey in the *Pinus pinaster* plantations established on the stabilized sands near the town, and extending as far as Monte Gordo. In spring, a prominent composite there is the yellow-flowered *Andryala integrifolia* (Plate 25).

Cliffs

Some of the sand-dune species besides plants restricted to the rocks are found on the Algarve cliffs and, like that of the dunes, cliff vegetation grades into the *matos* shrublands typical of the coastal region. Some of these cliffs are topped by sand as at Monte Clérigo and Bordeira (on shales) and at Cape St. Vincent (limestone).

On the steep slopes, sandstone, shale, and limestone, the grey–green leaved mat-forming composite with yellow heads of flowers, *Asteriscus maritimus* (Plate 26) is common, its old heads dried hard and grey. Frequently seen too are sea-beet, *Beta vulgaris* ssp. *maritima* (Chenopodiaceae, Plate 27), and the samphire 'Crithmum maritimum' but, especially, the intricately spiny *Lycium intricatum* (Solanaceae, Plate 28) with minute sweetly scented flowers, and the rosettes of *Plantago coronopus* (Plantaginaceae). The gigantic *Agave americana* (*pita*, Plate 29) is frequently naturalized and also, as at Praia da D. Ana (Lagos), is a large species of aloe, *Aloe vera*. This plant is the commercial source of bitter aloes used in many cosmetics. In the Algarve, its latex is traditionally considered to have antibiotic properties and is used for dressing external wounds.

At Praia da D. Ana, though now (1991) bulldozed, the cliff-top vegetation comprises a shrubby assemblage: *Atriplex halimus*, naturalized olives, *Suaeda vera*, *Chamaerops humilis*, (*palmeira anã, palmeira das vassouras*), *Osyris quadripartita*, *Thymus camphoratus* (Plate 30), *Prasium majus* (Labiatae), *Asparagus* spp., and species of *Lycium*, with the thorny scrambling liane *Smilax aspera* over all. There are also rather few herbs: sea campion, *Silene maritima* (Caryophyllaceae), the scrambling weedy

Figure 54 Sandstone cliffs, *O Pinhão*, Lagos.

Figure 55 Limestone cliffs at Cape St. Vincent.

Figure 56 Shale cliffs at Amoreira.

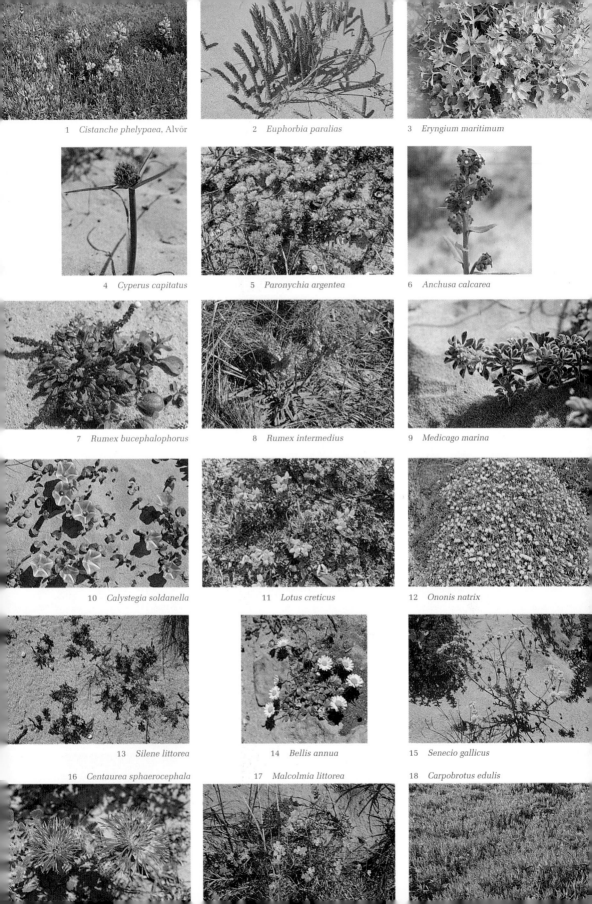

1   *Cistanche phelypaea*, Alvôr

2   *Euphorbia paralias*

3   *Eryngium maritimum*

4   *Cyperus capitatus*

5   *Paronychia argentea*

6   *Anchusa calcarea*

7   *Rumex bucephalophorus*

8   *Rumex intermedius*

9   *Medicago marina*

10   *Calystegia soldanella*

11   *Lotus creticus*

12   *Ononis natrix*

13   *Silene littorea*

14   *Bellis annua*

15   *Senecio gallicus*

16   *Centaurea sphaerocephala*

17   *Malcolmia littorea*

18   *Carpobrotus edulis*

19 *Carpobrotus edulis*, yellow form

20 *Carpobrotus edulis*, pink form

21 *Drosanthemum* sp.

24 *Retama monosperma*

22 *Armeria pungens*

23 *Arctotheca calendula*

25 *Andryala integrifolia*

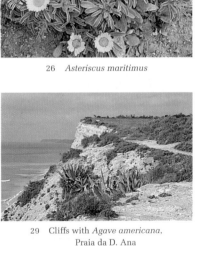

26 *Asteriscus maritimus*

27 *Beta vulgaris* ssp. *maritima*

28 *Lycium intricatum*

29 Cliffs with *Agave americana*, Praia da D. Ana

30 *Thymus camphoratus*

33 *Arisarum simorrhinum* in flower

31 *Phagnalon saxatile*

32 *Ruta angustifolia*

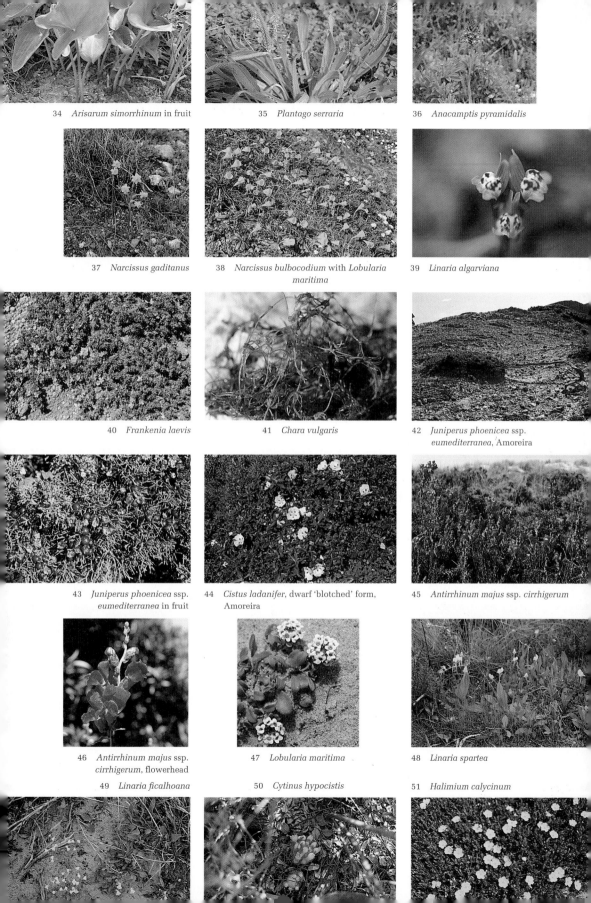

34   *Arisarum simorrhinum* in fruit

35   *Plantago serraria*

36   *Anacamptis pyramidalis*

37   *Narcissus gaditanus*

38   *Narcissus bulbocodium* with *Lobularia maritima*

39   *Linaria algarviana*

40   *Frankenia laevis*

41   *Chara vulgaris*

42   *Juniperus phoenicea* ssp. *eumediterranea*, Amoreira

43   *Juniperus phoenicea* ssp. *eumediterranea* in fruit

44   *Cistus ladanifer*, dwarf 'blotched' form, Amoreira

45   *Antirrhinum majus* ssp. *cirrhigerum*

46   *Antirrhinum majus* ssp. *cirrhigerum*, flowerhead

47   *Lobularia maritima*

48   *Linaria spartea*

49   *Linaria ficalhoana*

50   *Cytinus hypocistis*

51   *Halimium calycinum*

52   *Ephedra fragilis*

53   *Allium subvillosum*

54   *Dipcadi serotinum*

55   *Leucojum trichophyllum*

56   *Drimia maritima* flowering in autumn

57   *Silene colorata*

58   *Astragalus lusitanicus*

59   *Lupinus luteus*

60   *Bartsia trixago*

61   *Salsola brevifolia*

62   *Coronilla valentina*

63   *Astragalus tragacantha* ssp. *vicentinus*

64   *Stauracanthus genistoides*

65   *Daucus halophilus*

66   *Armeria pungens*

67   *Corema album*

68   *Bellevalia hackelii*

69   *Cistus ladanifer* ssp. *sulcatus*

70 *Prasium majus*

71 *Helianthemum origanifolium*

72 *Biscutella vincentina*

73 *Calendula suffruticosa* ssp. *suffruticosa*

74 *Calendula suffruticosa* ssp. *tomentosa*

75 *Cerinthe gymnandra*

76 *Euphorbia segetalis*

77 *Cuscuta* sp.

78 *Orobanche* sp. at Cape St. Vincent

79 *Orobanche* sp. at Cape St. Vincent

80 *Anagallis monelli*

81 *Romulea bulbocodium*

82 *Lithodora prostrata*

83 *Retama monosperma*

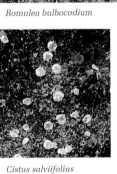
84 *Cistus salviifolius*

85 *Erica umbellata*

86 *Genista hirsuta*

87 *Quercus coccifera*

88   *Lavandula stoechas*   89   *Scilla monophyllos*

90   *Asphodelus ramosus*   91   *Fritillaria lusitanica*   92   *Serapias lingua*   93   *Gladiolus illyricus*

94   *Centaurium maritimum*   95   *Tolpis barbata*

96   *Thymelaea hirsuta*   97   *Halimium halimiifolium*   98   *Campanula lusitanica*

99   *Cistus ladanifer*   100   *Gynandriris sisyrinchium*   101   *Iris xiphium*

102   *Allium roseum*   103   *Cachrys trifida*   104   *Cistus albidus*

105   *Phillyrea angustifolia*

106   *Phlomis purpurea*

107   *Prasium majus*

108   *Vinca difformis*

109   *Centaurium erythraea*

110   *Iberis linifolia*

111   *Omphalodes linifolia*

112   *Romulea bulbocodium*

113   *Anemone palmata*

114   *Tulipa sylvestris* ssp. *australis*

115   *Chamaerops humilis* in flower

116   *Chamaerops humilis* fruits
and flowerbuds

117   *Sideritis linearifolia*

118   *Aceras anthropophorum*

119   *Narcissus gaditanus*

120   *Paeonia broteroi*

121   *Aristolochia boetica*

122   *Doronicum plantagineum*

123 *Gladiolus italicus*

124 *Gladiolus italicus*

126 *Asphodelus fistulosus*

125 *Iris planifolia*

127 *Orchis italica*

128 *Orchis morio*

129 *Ophrys scolopax*

130 *Ophrys tenthredinifera*

131 *Ophrys fusca*

132 *Ophrys lutea*

133 *Ophrys lutea*, lip of flower

134 *Ophrys bombyliflora*

135 *Ophrys vernixia*

136 *Foeniculum vulgare*

137 *Myoporum tenuifolium*
     hedge near Sagres

138 *Myoporum tenuifolium*

139 *Cydonia oblonga*

140 Oatfield weeds

141 *Chrysanthemum segetum*

142 *Chrysanthemum coronarium*

143 *Echium plantagineum*

144 *Silybum marianum*

145 Roadside weeds, mainly *Galactites tomentosa*

146 *Galactites tomentosa*

147 *Parentucellia viscosa*

148 *Bituminaria bituminosa*

149 *Lathyrus ochrus*

150 *Melilotus indica*

151 *Scorpiurus muricatus*

152 *Trifolium angustifolium*

153 *Trifolium stellatum*

154 *Convolvulus althaeoides*

155 *Convolvulus tricolor*

156 *Nigella damascena*

157 *Briza maxima*

158 *Geropogon hybridus*

159 *Papaver somniferum*

160 *Borago officinalis*

161 *Ecballium elaterium*

162 *Scilla peruviana*

163 *Muscari comosum*

164 *Narcissus papyraceus*

165 *Narcissus papyraceus*

166 *Genista triacanthos*

167 *Lavandula viridis*

168 *Coronilla valentina*

169 Cork oak flowers

170 *Cytisus scoparius*

171 *Euphorbia monchiquensis*

172 *Genista tridentata*

173 *Viburnum tinus*

174 *Cistus populifolius*

175 *Cistus crispus*

178 *Quercus canariensis* flowers

179 *Arbutus unedo* flowers and fruit

183 *Rhododendron ponticum* ssp. *baeticum*

186 Chestnut flowers

189 *Halimium calycinum*

176 *Paeonia broteroi*

177 *Quercus canariensis*

180 *Eucalyptus camaldulensis* flowers

181 *Eucalyptus globulus* flower

182 *Rhododendron ponticum* ssp. *baeticum*

184 *Romulea bulbocodium*

185 *Euphorbia characias*

187 *Quercus lusitanica* with gall

190 *Centranthus calcitrapae*   191 *Sedum forsterianum*

188 *Erica australis*

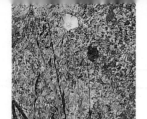

192   *Tuberaria lignosa*

193   *Drosophyllum lusitanicum*

194   *Limodorum abortivum*

195   *Arundo donax* along the Guadiana near Alcoutim

196   *Arundo donax* in flower

197   *Rosa canina* near Arão

198   *Iris pseudacorus*

199   *Baldellia ranunculoides*

200   *Alisma plantago-aquatica*

201   *Narcissus bulbocodium* at Lagoa Funda

202   *Narcissus bulbocodium* at Lagoa Funda

203   *Cotula coronopifolia* at Lagoa Funda (*Pinus pinaster* beyond)

204   *Cotula coronopifolia*

205   *Linaria amethystea*

206   *Ranunculus peltatus*

207   *Ranunculus peltatus*

208   *Serapias cordigera*

Convolvulus althaeoides (Convolvulaceae), *Phagnalon saxatile* (Compositae, Plate 31), *Ruta angustifolia* (Plate 32), *Arisarum simorrhinum* ('*A. vulgare*', Araceae, Plates 33 and 34), its heavy and knobbly fruiting head curving over to ground level as the leaves wither, and *Plantago serraria* (Plate 35). Orchids common in the disturbed grassland nearby include *Ophrys lutea*, *O. vernixia*, *O. apifera*, and the pyramidal orchid *Anacamptis pyramidalis* (Plate 36).

Near Salema, the tiny *Narcissus gaditanus* (Plate 37), named after the Latin word for Cadiz, is to be found, and the steep, grassy shale cliffs at the southern edge of Praia do Castelejo are covered in March with the flowers of *N. bulbocodium* (Plate 38) and *Romulea bulbocodium*. Amongst them are the sparkling flowers of the tiny *Linaria algarviana* (Plate 39), an Algarve endemic, and, on the rocks near the sea, *Frankenia laevis* (Frankeniacae, Plate 40) and *Inula crithmoides* (Compositae). Further up the slope is the woody composite *Centaurea vincentina* and the stream nearby is notable for a stonewort, *Chara vulgaris* (Plate 41).

On blown-out cliff-top dune areas over shales at Praia da Amoreira, there are remarkable 'moonscapes' caused by concentrations formed around the root-systems of long-disappeared shrubs. In areas still covered with vegetation, there are the grey of *Helichrysum picardii* and the bright green of *Ononis natrix*. In the highly disturbed dune systems at Alvôr, Oxford University students have found that the *Ononis* lives for only a few years. At Amoreira, it can tolerate more exposed areas than can *H. picardii*, which survives in such places only in close association with *Ononis*.

On the steep slopes at Amoreira, with the cliffs exposed to the full effect of the sea and prevailing winds, the only large plant is *Juniperus phoenicea* ssp. *eumediterranea* (Plates 42 and 43), some of the specimens over 2 m in length with bared root-systems appearing to be precariously attached to the surface, the mangy foliage acting as a trap for organic material in which other plants become established, notably the inconspicuous green-flowered orchid *Gennaria diphylla*. In less exposed positions, the juniper grows with the coastal dwarf form of *Cistus ladanifer* (Plate 44, see below for discussion) and in areas with larger amounts of sandy soil, the rose-scented *Antirrhinum majus* (Scrophulariaceae, Plates 45 and 46) pokes up through the juniper bushes. This is the ssp. *cirrhigerum*, which has pink–red flowers and upper leaves curling back and acting as grapples, holding the heavy inflorescences in the juniper canopy. With increasing amounts of sand, *Corema album* becomes common, as do *Thymus camphoratus* and *Daphne gnidium*. Between the bushes are the short-lived *Lobularia maritima* (alyssum, Cruciferae, Plates 38 and 47), familiar as a summer-bedding plant in northern Europe, and the minute *Bellis annua* (Compositae), a tiny daisy, and the toadflaxes *Linaria spartea* (Plate 48) growing to 20 cm in height with yellow flowers and the minute yellow-flowered *L. ficalhoana* (Plate 49), a Portuguese endemic, less than half as tall.

On the south-facing slopes at Amoreira, there is a rich flora grading into the typical *matos* of the region, conspicuous being the grey-leaved *Halimium halimiifolium*, often with the parasitic *Cytinus hypocistis* (Plate 50) growing from its roots, *Cistus salviifolius*, *Dorycnium hirsutum* (Leguminosae), *Halimium calycinum* (*H. commutatum*, Plate 51), *Ulex argenteus*, *Lavandula stoechas*, *Ephedra fragilis* (Plate 52) and, in the bare patches between bushes, the geophytes *Allium subvillosum* (Plate 53), *Dipcadi serotinum* (Plate 54), *Ophrys tenthredinifera*, and *Leucojum*

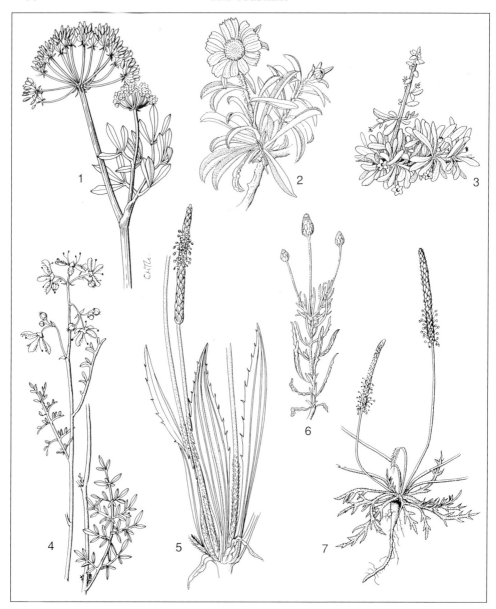

Figure 57   Algarve cliff plants. *Plantas de falésia*. 1. *Crithmum maritimum*; 2. *Asteriscus maritimus*; 3. *Lycium intricatum*; 4. *Ruta angustifolia*; 5. *Plantago serraria*; 6. *Phagnalon saxatile*; 7. *Plantago coronopus*.

trichophyllum (Plate 55), a snowflake, *Muscari neglectum*, a grape-hyacinth as well as the ubiquitous sea-squill *Drimia maritima* (Plate 56), a medicinal plant whose bulbs were formerly exported for diuretic production. Also there are the short-lived toadflaxes, as well as *Erodium aethiopicum* (Geraniaceae), *Tuberaria guttata* (Cistaceae), *Silene colorata* (Caryophyllaceae, Plate 57), *Astragalus lusitanicus*

Figure 58   Cliffs at Castelejo with *Narcissus bulbocodium*.

Figure 59   Concretions around old roots, Amoreira.

Figure 60   Cliff-top sand vegetation, Monte Clérigo.

Figure 61   Plant communities at Monte Clérigo, with *Astragalus tragacantha* ssp. *vicentinus (A. massiliensis)* in foreground.

(Leguminosae, Plate 58), which is toxic to stock, *Lupinus angustifolius* and *L. luteus* (Leguminosae, Plate 59), and the rapid-growing hemiparasite *Bartsia trixago* (Plate 60), producing its statuesque inflorescences of yellow flowers late in April. There is similar vegetation all along the coast from Odeceixe, where *Halimium verticillatum* is to be found, and the vegetation is encroaching again over uncompleted developments, as far as the limestones of the Cape St. Vincent area.

Cape St Vincent

Although many of the sites so far mentioned fall within the protected area of the south-west of the country (see Chapter 8, *Área Protegida da Costa Vicentina e Sudoeste Alentejano*), the Cape St Vincent area is the most renowned, both historically and as a classic site for biologists. The Sagres peninsula has steep rocks on all seaward sides, a table-land extension of the *Barrocal* vegetation of the central part of the Algarve (see Chapter 3) and sand systems derived from material blown up on to this and thus resembling those on shales at, for example, Praia da Amoreira to the north.

The whole *Promontorium Sacrum*, including both Sagres and Cape St. Vincent, was studied in detail in 1938–40 by Rothmaler. He recognized a number of plant-communities from the *Ammophila*-dominated sands in Martinhal harbour, through areas notable for *Limonium* and *Ononis* species, to the summit assemblages characterized by *Juniperus* and *Cistus*. This, the western coast with a zone typified by *Astragalus* to the east and south, borders on vegetation characterized by *Corema album* on the sandy areas and one characterized by *Ulex*, with patches of *Quercus coccifera* (*carrasco*) and *Pistacia lentiscus* (*aroeira*) further inland. Here also are the last sites in Portugal for the woody *Viola arborescens*.

Rothmaler's account covered the whole region, but a more recent study has analysed the Cape St. Vincent area itself in much greater detail as a 'baseline' for future studies of change due to human activities. It was possible to recognize certain characteristic groupings which are clear to the inquisitive eye in what might, at first sight, appear to be a homogeneous area. On the south-facing cliff-tops, the assemblage is notable for herbs with the wiry shrub *Salsola brevifolia* (Plate 61), a community subject to great exposure and early drought. The cliff slopes have more shrubby vegetation with abundant, sweet-smelling *Coronilla valentina* (Leguminosae, Plate 62), the 'cucumber plant' (from the taste of its leaves), *Sanguisorba minor* (Rosaceae), and the foetid *Ruta angustifolia*. The *Astragalus tragacantha* ssp. *vicentinus* (Plate 63) community of Rothmaler is notable also for the prickly hedgehog plants of the gorse-like *Stauracanthus genistoides* (Plate 64) with yellow flowers and opposite spiny leaves (*Ulex* has spirally arranged leaves) in an area on the north-facing slopes subject to fierce winds from the sea. *Carpobrotus edulis* and *Lotus creticus* are found on cliff-tops. *Juniperus* and *Teucrium luteum* are associated on a sheltered westerly slope, the *Limonium* community of Rothmaler being found on the most exposed northerly cliffs notable for *L. ferulaceum*, *L. ovalifolium*, and the carrot *Daucus halophilus* (Umbelliferae, Plate 65). Other recognizable groupings include a *Juniperus–Quercus coccifera* area on *terra rossa* (clayey and coloured by iron oxides) soils on the

Figure 62   Vegetation at Cape St. Vincent.

southern edge of the Cape. These soils in the centre of the peninsula, where they are deeper, have a vegetation typically with *Lavandula stoechas* and *Armeria pungens* (Plate 66), the latter with *Corema album* (Plate 67) characterizing the sand-dune communities.

Common geophytes on the plateau include *Narcissus bulbocodium*, *Allium subvillosum*, *Drimia maritima* with bulbs up to 2 kg close to the surface with a scape to 1 m, the terminal 60 cm of which bears about 200 flowers when the plant is leafless in September, the grape-hyacinth-like *Bellevalia hackelii* (Plate 68, the corolla-lobes not constricted like those in true grape hyacinths, *Muscari* spp.), and *Arisarum simorrhinum*. The major shrubs of the area are the sticky-leaved *Cistus ladanifer*, here in a dwarf form (ssp. *sulcatus*) with 'unblotched' petals, once known as *C. palhinhae* (Plate 69), but probably only a coastal ecotype which extends north and east from the Cape, sometimes (as at Praia da Amoreira) with 'blotched' petals, the 'blotches' being of different sizes on different plants. 'Unblotched' tall specimens appear sporadically in 'blotched' populations elsewhere. The shrubs appear to be short-lived (up to 10 years), with most of the plants at Cape St. Vincent certainly much younger. Despite their air of antiquity, the creeping forms of *Juniperus phoenicea* there are rarely over 40 years old, the bulk of the population being perhaps 10–25 years of age. The fruits develop at the end of their second year, but seeds germinate slowly and with difficulty, improved by passage through an animal, especially a bird. *Juniperus* shows a rapid increase in area with age in the first ten or so years, and then stabilizes. *Cistus* cycles much faster than *Juniperus*, which accumulates soil by trapping wind-blown debris and sand.

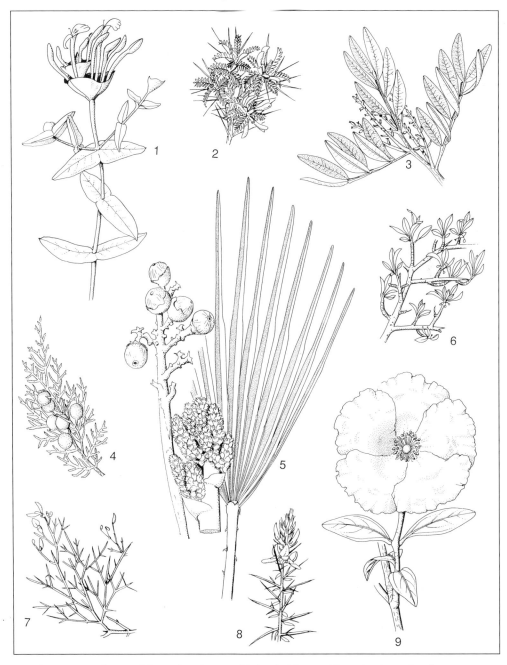

Figure 63 Plants of Cape St. Vincent *(Cabo S. Vicente)* I. 1. *Lonicera implexa*; 2. *Astragalus tragacantha* ssp. *vicentinus*; 3. *Pistacia lentiscus*; 4. *Juniperus phoenicea*; 5. *Chamaerops humilis*; 6. *Rhamnus lycioides*; 7. *Stauracanthus genistoides*; 8. *Genista hirsuta*; 9. *Cistus ladanifer* ssp. *sulcatus*.

Figure 64   Plants of Cape St. Vincent (*Cabo S. Vicente*) II. 1. *Rubia peregrina*; 2. *Silene colorata*; 3. *Viola arborescens*; 4. *Cerinthe gymnandra*; 5. *Biscutella vincentina*; 6. *Euphorbia segatalis*; 7. *Lavandula stoechas*; 8. *Romulea bulbocodium*; 9. *Armeria pungens*; 10. *Prasium majus*; 11. *Iberis contracta* ssp. *welwitschii*; 12. *Helianthemum origanifolium*; 13. *Anagallis monelli*.

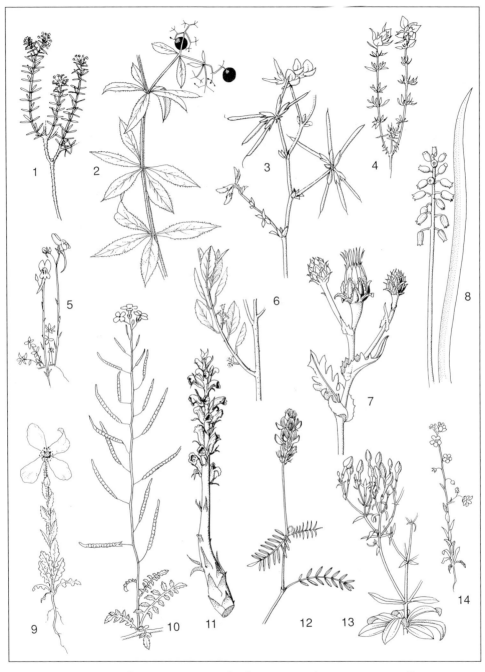

Figure 65 Plants of Cape St. Vincent (*Cabo S. Vicente*) III. 1. *Corema album*; 2. *Rubia peregrina* (fruit); 3. *Lotus creticus*; 4. *Thymus camphoratus*; 5. *Linaria algarviana*; 6. *Rhamnus alaternus*; 7. *Reichardia gaditana*; 8. *Bellevalia hackelii*; 9. *Papaver somniferum*; 10. *Diplotaxis vicentina*; 11. *Orobanche minor*; 12. *Onobrychis humilis*; 13. *Centaurium erythraea*; 14. *Omphalodes linifolia*.

Nevertheless, in these communities, there would appear to have been very few major fires in the last few decades.

Other shrubs at Cape St. Vincent include *Chamaerops humilis*, *Cistus monspeliensis*, *Cistus salviifolius*, *Thymus camphoratus*, with large leafy bracts beneath the flower heads, *Daphne gnidium*, flowering later in the year, *Rhamnus alaternus*, *R. lycioides*, and *Phillyrea angustifolia*. Subshrubs include the woody labiate *Prasium majus* (Plate 70) with white flowers, glossy leaves, and succulent fruits unlike most of its family, the hemiparasitic *Osyris quadripartita*, *Helianthemum origanifolium* (Plate 71), grading into woody-based herbs such as *Astragalus lusitanicus*, with its baggy pods and dirty-cream flowers, often with red calyces, *Anthyllis vulneraria*, *Biscutella vincentina* (Plate 72), with its figure-of-eight double fruits, *Diplotaxis vicentina*, both cruciferous plants restricted to southwestern Portugal, *Calendula suffruticosa* (Plate 73) and its hairier subspecies *tomentosa* (Plate 74), *Cerinthe gymnandra* (Plate 75), *Onobrychis humilis*, *Silene latifolia* (white campion), and *Ruta angustifolia*.

In clear, largely sandy, areas there are a number of annual or other short-lived plants such as the daisy, *Bellis annua*, *Centaurium erythraea* (Gentianaceae), *Euphorbia segetalis* (Plate 76), *Iberis contracta* ssp. *welwitschii*, *I. crenata* (*I. sampaiana*), *Lobularia maritima*, *Silene colorata*, *Mercurialis ambigua*, *Erodium aethiopicum*, the minute *Valantia muralis*, the Algarve endemic *Linaria algarviana*, *Paronychia argentea*, and *Rumex bucephalophorus*.

Succulents include *Umblilicus rupestris*, the tough *Sedum nicaeense*, and lianes, as well as the dodder *Cuscuta* sp. (Convolvulaceae, Plate 77), woody *Asparagus* spp., *Smilax aspera*, and *Lonicera implexa*. Of orchids, the green-flowered *Gennaria diphylla* and *Ophrys fusca* are conspicuous in spring, as are the parasitic *Cytinus hypocistis* ssp. *macranthus*, and *Orobanche minor*, the robust tufts of the sedge *Schoenus nigricans* and the grass *Stipa gigantea* persisting throughout the year. By May, the vegetation has a distinctly 'drier' aspect but species of *Orobanche* (Plates 78 and 79) are prominent and later-flowering plants such as *Linum strictum* can be seen. At this time, the highly disturbed Sagres Peninsula is bright with the yellow heads of *Asteriscus maritimus*.

Along the equally disturbed roadsides, mats of *Anagallis monelli* (Primulaceae, Plate 80) are very colourful, while notable in the sand-dune areas are the shrubby *Corema album*, *Halimium halimiifolium* and *H. calycinum*, the crocus-like *Romulea bulbocodium* (Iridaceae, Plate 81), *Lithodora prostrata* (Plate 82), and *Rosmarinus officinalis*, the tiny annual *Ionopsidium acaule*, (Cruciferae), sometimes grown in paving in northern Europe, with the prickly thistle-like *Centaurea sphaerocephala*, and the hare's-foot grass *Lagurus ovatus*.

3

The lowlands

The low-lying southern Algarve (*Litoral*) has long been affected by humans and much of the land which is not in production is rapidly being developed for building associated with the tourist industry. There are surviving seminatural communities of plants, maintained by land management régimes and often kept in check by fire, trampling, and grazing.

Near Faro and elsewhere along the coast are stands of umbrella pine (or 'stone' pine), *Pinus pinea* (*pinheiro manso* — *el pino piñonero* in Spain). This tree is widely planted, and is largely found on the coast; it is centred on Iberia but is known as far east as the eastern Black Sea. Its fleshy seeds are the pine nuts (*pignons*) of the kitchen. The cones ripen in February and the nuts (*pinhões*) are shaken down with long poles of 'bamboo' cane (*canas*, *Arundo donax*, Gramineae). In the Algarve there are two ways of removing the nuts which are tightly enclosed within a nearly spherical cone. In one method, the cones are crushed between two stones, in a hammer and anvil fashion. More commonly, the cones are put on the stone base of a hot oven of traditional design, and they split open, thus making it easy to extract the nuts. A day's work by an experienced person can yield more than 10 l of them. If not to be eaten fresh, but intended for storage or further processing, they are winnowed and then dried in the sun for a few days. Most are sold as

Figure 66 *Pinus pinea* woodland near Montenegro.

Figure 67   *Pinus pinea* cone and *pignons*.

harvested, many being used as topping for small cakes. Others (*pinhõ es azucerada*), like almonds, are coated with sugar and sold at Easter (*Páscoa*), though this cottage industry has declined in the Algarve. The extraction of pine kernels is not now significant in the province, although a major activity remains in parts of Alentejo, to the north, for example around Alcácer do Sal, the national annual production (1989) being 1360 t. The seeds are eaten by birds, some of which store them, thus dispersing the pine.

Other species of pine, including the introduced and frost-sensitive *P. pinaster*, which can be seen in large tracts near Vila Real de Santo António, have winged seeds which are dispersed by the wind. In those groves there is a thick undergrowth of *Retama monosperma* (Leguminosae, Plate 83), as in the dunes beyond. Around Montenegro, west of Faro, pine woodlands have well-grown forms of *Chamaerops humilis* (dwarf fan palm, *palmeira anã*), with distinct trunks, and the steep dunes of shingle bear *P. pinea* with a thick *mato* cover of *Cistus salviifolius* (*sargaço*, Plate 84), *C. crispus*, *Erica umbellata* (Plate 85), *Genista hirsuta* (Plate 86), *Quercus coccifera* (Plate 87), *Lavandula stoechas* (Plate 88), *Chamaerops humilis*, *Halimium calycinum* (*verdinho*), *Rosmarinus officinalis*, *Armeria pungens*, *Daphne gnidium*, and *Lithodora prostrata*. Conspicuous geophytes in spring include *Scilla monophyllos* (Plate 89), *Asphodelus ramosus* (*gamão*, Plate 90), *Fritillaria lusitanica* (Plate 91), *Ornithogalum broteroi* (Liliaceae), and the orchids *Serapias lingua* (Plate 92) and other species, with *Gladiolus illyricus* (Plate 93) flowering later.

Short-lived plants on tracks and in clearings include *Centaurium maritimum* (Plate 94), *Tolpis barbata* (Plate 95), *Tuberaria guttata*, and *Thymelaea hirsuta* (Plate 96). In poorly drained areas with *Stauracanthus* are to be found *Tuberaria major*, an Algarve endemic, and *Thymus camphoratus*, discovered by the French botanist, J.P. le Tournefort when he travelled in 1689 round the Algarve from Mértola to Castro Marim, Tavira, Faro, Portimão, Lagos, Cape St. Vincent, Vila do Bispo, Aljezur, and Odeceixe. In the wettest parts is 'royal fern', *Osmunda regalis*, and, in the driest, are *Halimium halimiifolium* (Plate 97) and *Thymus mastichina*,

Figure 68 *Matos* plants I. 1. *Sideritis linearifolia*; 2. *Arenaria montana*; 3. *Vinca difformis*; 4. *Rosmarinus officinalis*; 5. *Jasminum fruticans*; 6. *Saxifraga granulata*; 7. *Cistus ladanifer*; 8. *Cistus salviifolius*; 9. *Pistacia lentiscus*.

Figure 69 *Matos* plants II. 1. *Ornithogalum umbellatum*; 2. *Scilla monophyllos*; 3. *Tulipa sylvestris* ssp. *australis*; 4. *Dipcadi serotinum*; 5. *Fritillaria lusitanica*; 6. *Narcissus bulbocodium*; 7. *N. gaditanus*; 8. *Hyacinthoides hispanica*; 9. *Allium roseum*.

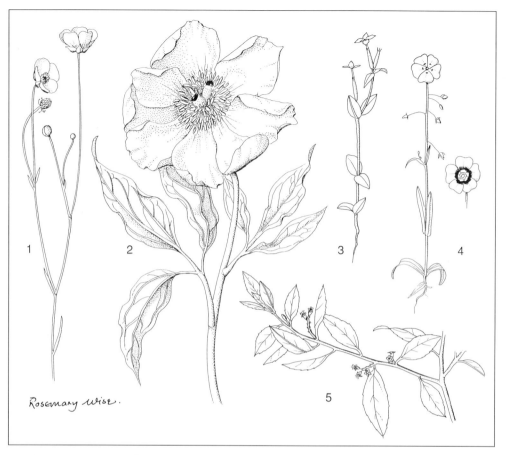

Figure 70 *Matos* plants III. 1. *Ranunculus gramineus*; 2. *Paeonia broteroi*;
3. *Centaurium maritimum*; 4. *Tuberaria guttata*; 5. *Rhamnus alaternus*.

with the short-lived *Malcolmia lacera*, *Campanula lusitanica* (Plate 98), and
Verbena officinalis.

The *mato* also includes large numbers of bushes of *Cistus ladanifer* (*cisto
ladanifero*, *lábdano*, Plate 99) which in the nineteenth century was commercially
significant in the Algarve, as it was in Crete and Cyprus. The resin in the sticky leaves
was distilled, and used as a basis for medicaments and perfumes, often by extraction
with ethanol. This extract must not be confused with '*laudanum*', which is a tincture
of opium, another former product of the *mato* (p. 92). After harvesting of the leaves of
C. ladanifer, the bushes were fired, producing new shoots the following year. The
flesh of partridges (*perdiz*) raised in the *matos* tastes of *C. ladanifer*.

Although some species which cannot tolerate lime, for example *Erica umbellata*,
are not to be seen on the *Barrocal*, none the less there is great species-richness in the
matos. In the south-west of the Algarve, the vegetation is much damaged, but relict

Figure 71 *Matos* plants IV. 1. *Aristolochia baetica*; 2. *Thapsia villosa*; 3. *Iris xiphium*; 4. *Tamus communis*; 5. *Erica umbellata*; 6. *Astragalus lusitanicus*; 7. *Gladiolus illyricus; 8. Daphne gnidium.*

patches in developed areas along the road to Sagres from Lagos in March are bright with *Gynandriris sisyrinchium* (Plate 100), which does particularly well on roadside verges. Other plants found in the relict *matos* are the orchid *Serapias parviflora*, *Iris xiphium* (the 'Spanish Iris' of gardens, Plate 101), *Gladiolus illyricus*, *Bellevalia hackelii*, *Allium roseum* (Plate 102) with its flowers scented like pinks, *Jasminum fruticans*, and *Cachrys trifida* (Umbelliferae, Plate 103).

The *Barrocal*

Where more species have survived, as in parts of the *Barrocal* for example at Salema, many of them grow more luxuriantly than they do at Cape St. Vincent, making a thicket to a height of 2 m, compared with the knee-high formations of similar species at the Cape. *Cistus ladanifer* overtops the rest, but the bee-pollinated *Lithodora prostrata*, *C. salviifolius*, and *C. albidus* (Plate 104), *Rosmarinus officinalis* — whose scented flowers (*flor de alecrim*) were picked until the early 1800s for export to Hamburg and Gdansk (Danzig) to be used in the manufacture of perfumes, *Quercus coccifera*, and *Arbutus unedo* grow to 7 m in height. *Jasminum fruticans*, *Phillyrea angustifolia* (Plate 105), *Rhamnus lycioides*, *R. alaternus*, *Juniperus phoenicea*, and *Daphne gnidium* can all exceed 1 m here. Other shrubby species include *Lavandula stoechas* and *Phlomis purpurea* (Labiatae, Plate 106), *Halimium calycinum*, and *Prasium majus* (Plate 107).

The more herbaceous plants include the scrambling *Arenaria montana* (Caryophyllaceae), *Vinca difformis* (a periwinkle, Apocynaceae, Plate 108), and *Astragalus lusitanicus*, whilst short-lived plants between the bushes include *Centaurium erythraea* (Gentianaceae, Plate 109), *Iberis linifolia* (Cruciferae, Plate 110), and *Omphalodes linifolia* (Boraginaceae, Plate 111). Growing up from rosettes of leaves are the umbellifers *Cachrys trifida* and *Thapsia villosa*, and the true geophytes include the local bluebell *Hyacinthoides (Scilla) hispanica* (Liliaceae), *Narcissus bulbocodium*, *N. gaditanus*, *Fritillaria lusitanica*, *Romulea bulbocodium* (Plate 112), *Valeriana tuberosa* (Valerianaceae), *Anemone palmata* (Ranunculaceae, Plate 113) with yellow or occasionally white daisy-like flowers, and *Tulipa sylvestris* ssp. *australis* (Plate 114) with yellow flowers, probably the only tulip truly native in Europe. Also there are the lianoid *Tamus communis* and the orchids *Gennaria diphylla*, *Ophrys lutea*, *O. bombyliflora*, *O. scolopax*, *O. apifera*, *O. vernixia*, and *O. tenthredinifera*, but most of these thrive better in more disturbed areas (see below). On drier, south-facing slopes are the palm *Chamaerops humilis* (Plates 115 and 116), its wedges of yellow flowers tucked deep into the heart of the plant near ground level and visited by beetles, the woody labiate *Sideritis linearifolia* (Plate 117), *Asperula hirsuta*, *Fumana thymifolia* (Cistaceae), and *Ruta angustifolia*. By early May, the *mato* here has a drier aspect and is less floriferous, but the shrubby labiates *Nepeta tuberosa* and *Teucrium haenseleri* are in flower, as are the short-lived *Centaurium maritimum*, *Linum strictum*, and *L. setaceum*.

The *Barrocal* extends from Sagres and Vila do Bispo in the west to a point just north-west of Tavira, and is species-rich. A recent study near Loulé provided three species new to the recorded Portuguese flora. These are *Euphorbia sulcata*, also known from nearby Andalucia, *Bunium bulbocastanum* (Umbelliferae), and *Crucianella latifolia* (Rubiaceae), the nearest known populations of these being in the Balearics. Amongst other striking plants of the *Barrocal* are species such as the man orchid *Aceras anthropophorum* (Plate 118), *Narcissus gaditanus* (Plate 119), *Sideritis linearifolia*, and *Stipa tenacissima*, which in Portugal are restricted to this habitat.

Figure 72   *Matos* on the *Barrocal*.

Figure 73   Heavily grazed *matos* on the *Barrocal*.

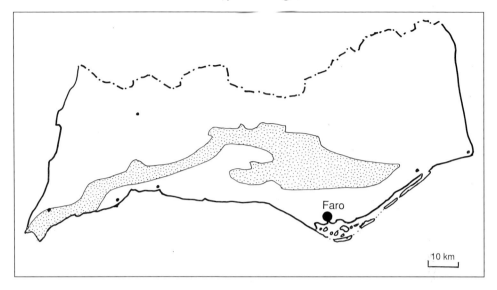

Figure 74   The limestone region. *O Barrocal Algarvio*. (Redrawn from Malato Beliz 1986.)

Making up low shrubland at the summit of the Rocha dos Soidos (altitude 467 m) north-east of Alte, are to be found *Juniperus phoenicea*, *Jasminum fruticans*, *Prasium majus*, *Chamaerops humilis*, *Rhamnus alaternus*, *R. lycioides*, *Rosmarinus officinalis*, *Cistus albidus*, and oleaster with sapling carobs. There is a rich flora of geophytes, heralded by *Narcissus gaditanus*, the beautiful pink-scented *Paeonia broteroi* (Plate 120) flowering later, *Hyacinthoides (Scilla) hispanica*, *Ornithogalum* sp., and a number of orchids including *Ophrys vernixia*, *Orchis morio*, and *Cephalanthera longifolia*, the scrambling 'flytrap' *Aristolochia boetica* (Plate 121), the parasitic *Cytinus hypocistis* (a relative of the Sumatran *Rafflesia arnoldii*, with flowers up to 1 m, the largest in the world), and a range of other plants including *Phlomis purpurea*, *Saxifraga granulata*, *Asplenium ceterach*, *Polypodium cambricum* and *Umbilicus rupestris* (*umbigo de Vénus*, Crassulaceae). Similar plants are found on the more disturbed Rocha da Pena nearby where the previously unrecorded *Doronicum plantagineum* (Compositae, Plate 122) produces its solitary capitula in spring.

Features of *mato* vegetation

Besides the geophytes and short-lived plants which avoid the summer heat, other plants of the *mato* have adaptations providing tolerance of the climate. Malacophyllous plants are soft-leaved species characteristic of various regions with seasonal rains alleviating drought. During such dry periods the leaves wilt but are able to withstand the stress and to recover after rain. Typical in the Algarve are species of *Cistus*, *Rosmarinus*, and *Thymus*, and other genera of their families (Cistaceae, Labiatae). When exposed to greater droughts, the leaves are shed, so reducing transpiration loss to that from stems and dormant buds. In southern Spain, *Rhamnus lycioides* (*espinheiro-preto*) becomes completely leafless during its flowering time in summer. On the other hand, there are sclerophylls, for example, oaks and olives with long-lived leaves, and the leafless 'switchplants' among the legumes and those like *Jasminum fruticans* with photosynthetic stems and small leaves.

The factors maintaining the *matos* include the felling of trees for building, fuel, and for the production of charcoal, grazing and browsing especially by goats, fires (intentional and otherwise), and agricultural clearance, the intensity and frequency of each of which in combination govern the overall appearance of the landscape. Elsewhere, the term *maquis* is used to describe evergreen formations greater than 1 m in height, and *garrigue* for that typically 50 cm, although the Spanish term *matorral* covers both. In the eastern Mediterranean at least, *maquis* species-richness increases with decreasing rainfall, but falls under extremely dry conditions. Similarly, grazing increases the diversity until severe over-grazing begins to reduce it.

Effects of fire

In Catalonia during the Spanish Civil War, George Orwell found that rosemary would burn once a fire was alight, but that *Quercus coccifera* was practically unburnable. When burnt, different species recover at different rates, *Cistus ladanifer* soon returning, *Erica arborea* by contrast being one of the slowest. Of the common trees in the Algarve, pines recover very slowly, and often not at all, but most planted eucalypts shoot from the base and along the trunk, and cork oak does so from the upper canopy; *Arbutus unedo* regenerates new growths, usually from the base, within one month of a severe burn. The frequency of fires thus affects both species composition and diversity.

The effect of fire on *maquis* plants has been studied in southern France, where many Algarve species are found. Six species there, including *Lonicera implexa* and *Daphne gnidium*, flowered in four months after a burn at the end of May. In the year following an autumn fire, 25 species were flowering or fruiting eight months after the burn. Almost all of these plants were herbaceous. The greatest number of species flowering

Figure 75 *Cistus ladanifer* sprouting after fire.

Figure 76   *Matos* regrowth after a major fire.

Figure 77   *Pinus pinaster* killed after fire.

for the first time after fire was during the year following the burn. Indeed 67 per cent of the French *maquis* flora flowers in the first year after a fire and this includes almost half the woody species, although flowers of *Quercus coccifera* are rare there, and *Cistus salviifolius* flowers only from sprouts, since the seedlings take some two or more years to bloom.

Those plants which tolerate fire can be considered 'passive', in that they withstand fire because of thick bark, for example cork oak, or have underground organs, for example bracken, or they are 'active', in that vegetative growth appears to be stimulated by fire, thereby producing shoots from underground stumps as in *Quercus coccifera*, *Juniperus phoenicea*, *Arbutus unedo*, *Erica arborea*, or seeds germinate after fire as in many Cistaceae, such as all the Algarve *Cistus* spp. and *Halimium halimiifolium*. The seed-coat, apparently the inner integument, is the barrier to germination by preventing water-uptake unless damaged.

Cistus monspeliensis can reproduce only by seeds, so if the interval between burns is less than four years, the species disappears from the affected site. Some species did not flower in a six-year interval between burns, and, of Algarve species, these included *Pistacia lentiscus*, *Rhamnus alaternus*, *Phillyrea angustifolia*, *Sedum nicaeense*, and a blackberry *Rubus ulmifolius* (*R. inermis*). *Pistacia lentiscus* was still unable to flower eight years after the fire. Flowering of most of the herbaceous perennials by contrast was promoted by fire, which may be due to the opening-up of the community. The species-richness increases as the annuals come into flower, but declines as they are crowded out. After fire in the *garrigue* of southern France, floristic richness is low during the first twelve months, reaching a maximum after three years, then diminishing with apparent stabilization when five years are past. Immediately after burning, there are increases in levels of nutrients, but further burnings do not repeat this since there is no time then for the build-up of biomass. However, nutrient levels remain higher in sites in *Quercus coccifera garrigue* burnt regularly in the autumn compared with those burnt in spring, although nitrogen levels drop as a result.

Nearer to the Algarve, at Doñana in Andalucia, there is a growth of herbs for four or five years after fire when the *mattoral* takes over. Of the *mattoral* species, *Chamaerops humilis* and *Daphne gnidium* are outstanding in their capacity to recover. Immediately after the fire, injured leaves as well as new ones are seen to be growing on the palm; although *Daphne gnidium* loses all its branches in the fire, it produces new shoots growing up to 70 cm in three months. *Erica australis* and *E. arborea*, by contrast, will only grow from such shoots if water is available, and bracken and *Drimia maritima* grow rapidly after rain and apparently greatly benefit from the increased nutrients and decreased competition following the fire. *Halimium halimiifolium*, *H. calycinum*, and *Cistus salviifolius* will not resprout there, but fire is followed by rapid and copious germination. *C. salviifolius* will not grow very vigorously, however, until there is some shelter from other *matorral* species.

Cistus spp. have been used to revegetate arid, erodable sites in southern Californian watersheds and are less flammable than the native *chapparal* species. On degraded sites in Spain, pure stands can persist for many years. *Cistus ladanifer* is the most flammable, for unlike *C. albidus*, *C. monspeliensis*, and *C. salviifolius*, which contain no monoterpenes, its leaves have 0.1–0.2 per cent of their fresh weight as α-pinene, which has a flash point of 32.8°C!

Effects of animals

Apart from fire, a second dominating ecological factor in the Mediterranean is grazing and browsing by animals. The original fauna of large animals has been greatly reduced

Figure 78   Grazing coastal *matos*.

Figure 79   Grazing grasslands replacing *matos*.

but it is estimated that today the region has some 270 million 'sheep equivalents' dependent on the Mediterranean vegetation, and that such pressure represents consumption of 1.5 t of dry matter per hectare per year. In 1976, Portugal had some 1 828 000 sheep and 350 000 goats, with 570 000 cattle in the Mediterranean zone. Of these animals, only goats have a metabolism that can cope with a diet of woody browse alone, and a pair of goats can give rise to a herd of 100 in five years, whereas sheep produce only 32 and cattle only 10 offspring in this time.

Goats and other animals favour certain plants over others, so that *Ceratonia siliqua*, *Erica arborea* and other heathers, olives, *Phillyrea angustifolia*, *Pistacia terebinthus*, *Quercus coccifera* and other oaks, *Rhamnus* spp., and *Viburnum tinus* are selectively browsed, whereas most Labiatae, many Umbelliferae and Compositae, many Leguminosae and Cruciferae, Euphorbiaceae, and almost all Liliaceae, are avoided altogether. Some of these plants which are rich in essential oils have been chosen by humans as flavourings for their foods, though many contain toxic alkaloids, mustard oils or other effective deterrents to herbivory. Clearly then, under different grazing and browsing régimes by various kinds of livestock, the vegetation will appear very different.

Although the most far-reaching effects of animals are due to grazing and browsing by introduced livestock, animals are intimately involved in the pollination of flowers and dispersal of seeds of *matos* plants.

Cistus

The flowers of *Cistus albidus* and *C. monspeliensis* usually last only one day, whilst *Cistus ladanifer* (as at Salema) often lasts two days or longer. A wide range of insects visits all three species, some beetles perhaps being important in the pollination of *C. monspeliensis* (larger hairy ones often being seen in *C. albidus*, though whether they do anything more than mate there is still unclear) with both honey and bumble bees being more important in *C. albidus* and *C. ladanifer*. Species of *Cistus* readily form hybrids, many of them important in horticulture, *C. monspeliensis* can only be a pollen parent but it crosses with *C. salviifolius*, to give *C. x florentinus*, and with *C. populifolius* to give *C. x nigricans*.

Table 1 Hybrids between *Cistus* spp. native to the Algarve

	albidus	crispus	monspeliensis	salviifolius	populifolius
albidus	*albidus*				
crispus	[1]	*crispus*			
monspeliensis	—	—	*monspeliensis*		
salviifolius	—	—	[2]	*salviifolius*	
populifolius	—	—	[3]	[4]	*populifolius*
ladanifer (including *palhinhae*)	[5]	—	[6]	[7]	[8]

[1] *C.* x *pulverulentus*
[2] *C.* x *florentinus*
[3] *C.* x *nigricans*
[4] *C.* x *corbarienis*

[5] *C.* 'Anne Palmer'
[6] *C.* x *loretii*
[7] *C.* x *verguinii*
[8] *C.* x *aguilari*

Figure 80 *Cistus ladanifer* ssp. *sulcatus* ('*C. palhinhae*').

Figure 81 *Cistus ladanifer* (tall 'unblotched' form).

Figure 82 *Cistus albidus.*

Figure 83 *Cistus salviifolius.*

Of the 15 possible hybrid combinations between the *Cistus* species found in the Algarve, seven are known in the wild, and an eighth (*C.* 'Anne Palmer') has been produced in cultivation. Hybrids are also formed with *Halimium* spp. and other members of Cistaceae. This shows that the pollinators are not confined to individual plant species over extended periods and that there are few genetic blocks to crossing.

Asparagus aphyllus

Insects can also greatly affect seed production. In Andalusia, *Asparagus aphyllus* flowers in July, but few fruits mature owing to the depredations of the larvae of the moth *Metachrostis dardouinii*. Generally, a second crop follows in August–September, but is also destroyed. A third crop in November–December is more successful, but in places where the insect is rare, the earlier fruitings are successful, so that regional variation in fruiting in *Asparagus* is associated with variation in past abundance or host-plant selection.

Osyris and *Pistacia*

Osyris quadripartita is an evergreen dioecious shrub of the Mediterranean. In southern Spain, the females grow from November until August, the males almost all the year round, starting one and a half months ahead of the females, shortly after the autumn rains. Female plants flower from March to September, males all the year round with a peak in May–June. Most flowering takes place in spring, and fruits develop in the dry summer, ripening at any time of the year, but the probability of setting ripe fruits steadily decreases from early to late season flowers owing to increasing rates of abortion, a reflection of resource limitations imposed by the increasingly desiccating season. There is therefore a major peak in winter and a minor one in spring. This behaviour is familiar in tropical rain-forest plants, suggesting that *O. quadripartita* is a part of the old evergreen tropical flora relict in the Mediterranean.

The hemiparasitic habit of *Osyris* may allow virtually continuous activity throughout the year in a strongly seasonal climate. The difference between males and females appears to be associated with the constraints of fruit maturation.

Osyris quadripartita is dispersed largely by a warbler, *Sylvia atricapilla* the blackcap, a winter visitor, and it parasitizes, in about 90 per cent of cases in southern Spain, other species of fleshy-seeded shrub, especially *Pistacia lentiscus*, also a dioecious species, and *Phillyrea angustifolia* or *Juniperus phoenicea* and *Corema*

Figure 84 Blackcap (*Sylvia atricapilla*). (Photograph: A Rocha Observatory)

Figure 85 *Pistacia lentiscus.*

album. These are more parasitized than would be expected from their relative abundance. The droppings of the bird have been found to contain mixtures of up to five different species of seeds, so that, in general, seeds of the parasite 'travel' in the gut with those of other species. The percentage occurrence together as 'gut-fellows' neatly correlates with the percentage combination of the species as mature plants. *P. lentiscus* females are parasitized more often than males, suggesting that the association is linked to the feeding-place immediately after ingesting *Osyris* fruits.

Pistacia lentiscus, the staple food at El Viso, Andalucia of the same warbler, although rich in lipids and protein, needs to be balanced by certain 'minor' fruits, important among which is *O. quadripartita*, though it is not too common overall. That females more often parasitize the *Pistacia* than do males means that the proximity to the *Pistacia* will ensure greater representation. In short, as a result of the food-selection of the warbler, the seeds are dispersed in multi-species packages. This explains the host-use by the parasite, thereby mirroring the gut-fellow proportions. In turn this produces a feedback between dispersal interaction and the structure of the habitat, strengthening the relationship with time between all three components. The *Pistacia* fruits are an important food resource for migrating frugivorous birds, though 83 per cent of seeds recorded from Andalusian plants were taken by just four bird species, to dispersal by which *Pistacia* seems adapted, though its origins are in the tropics where the avifauna is very different.

In the nineteenth century, *Pistacia lentiscus*, as well as *Phillyrea angustifolia*, was collected in the Algarve for resin. The resin was used both medicinally and in commercial manufacture of varnishes, uses known to the Romans. The mastic from the plants is said to have beneficial effects on the gums, and provides a good mouthwash. The branches of *P. angustifolia* were also used for making brooms. From sumach, *Rhus coriaria*, a shrub also in Anacardiaceae, comes the tanning material which gave the original red colour to both Cordovan and Moroccan leather, and was once a very conspicuous component in the *matorral* of the Sierra Morena (Spain). It was particularly well known in the mining centre of Guadalcanal (Seville). Sumach was cut in August, with the stems, leaves, and flowers being milled and taken to the tanneries. In the nineteenth century it was grown commercially in the Algarve, in much the same way as vines.

Smilax aspera

Smilax aspera berries may have one, two or three seeds, the single-seeded ones having the largest ratio of pulp to seed, thus offering more to a dispersal agent. However, numbers vary between individuals as well as populations, and seem to be correlated best with the number of competing fruiting plants, that is to say, competition for dispersal agents (birds) when fruits are in short supply. In those cases, birds select the more profitable fruits, the differences between which must be genetically controlled. In the *mattoral* of Andalucia there are indeed many potential competitors when the black berries of the *Smilax* are available between September and March. Other plants with such berries at the same time include *Osyris quadripartita*, *Pistacia lentiscus*, *Daphne gnidium*, *Myrtus communis* (myrtle, Myrtaceae), *Rubia peregrina*, *Viburnum tinus* (Caprifoliaceae), and *Ruscus aculeatus* ('butcher's broom', Liliaceae).

In southern Spain, the black fruits of *Rubus ulmifolius* (*R. inermis*) are dispersed mainly by migrant birds, however. Most birds are not exclusively frugivorous, and those in the eastern Mediterranean, when fed on fruit alone, lose weight in spite of increased rates of pulp intake. Surprisingly, this loss is greater in the more frugivorous birds. Most birds ingest more protein and energy than required but cannot digest the food, so that it has been suggested that the fruits may contain agents which reduce digestibility, and birds may then move away from sites rich in fruit to eat insects, thereby improving their diet and thus avoiding depositing seeds below the parent plants which would lead to competition between parent and offspring.

The relationship between the *mato* plants and their avian dispersal agents is such that those fruiting in summer, when the birds are short of water, have watery fruits, whilst those ripening in winter, when the birds need high-energy food, are rich in fats. Many migrant garden warblers in autumn feed extensively on figs which are considered important for the rapid deposition of migratory fat required for crossing the Sahara.

Roadsides and fields

Much of the lowland landscape has had even the *matos* removed. Instead there are grassy herbaceous communities, which are also found under the tree-crops typical of the Algarve. At a further stage, these are replaced by herbaceous crops, particularly beans (broad beans and 'French' beans), peas, and cereals (see Chapter 4). Firstly the 'wild' plants under such modified conditions will be considered and then the crops themselves in Chapters 4 and 5.

In spring the roadsides and cultivated areas are colourful with introduced exotics, the most common being the ubiquitous 'Bermuda Buttercup' (*trevo azédo*, *Oxalis pes-caprae*), from southern Africa. In the Mediterranean it was first recorded in Sicily in 1796, Malta in 1806, Libya in 1824, and Portugal in the following year. Introduced

Figure 86 *Oxalis pes-caprae.*

Figure 87 *Oxalis pes-caprae.*

Figure 88 *Oxalis pes-caprae*, partial 'double' form.

Figure 89 *Oxalis pes-caprae*, fully 'double' form.

Figure 90   *Iris* x *albicans.*

Figure 91   *Iris* x *albicans.*

Figure 92   *Pelargonium* at roadside
near Sagres.

as an ornamental, it is now established in many warm parts of the world and has
underground bulblets, by which it spreads rapidly in any disturbed soils. Indeed, seed
capsules were not recorded in Portugal until 1938, four years after the first report from

Figure 93 *Watsonia meriana* at roadside near Odeceixe.

Figure 94 *Zantedeschia aethiopica*. (Photograph P. Garton.)

Tunisia. Wild populations in Africa have flowers of three different types on different plants. These vary in style length, and seeds are produced only by crossing forms with different style-lengths. The Mediterranean (and Australian) populations are largely pentaploid and short-styled, and were apparently derived from crossing diploid and tetraploid forms in South Africa. Not infrequently in the Algarve, where *O. pes-caprae* flowers from November until April, there are forms with 'double' flowers, where the stamens are replaced by petals. Its distribution is facilitated by increased irrigation. The plant is toxic to stock because of its high levels of oxalic acid.

Deliberately introduced are the white irises so conspicuous along the roads near Loulé and on the EN125 between Lagoa and Faro. These are *Iris × albicans*, a natural hybrid which is related to the *Iris × germanica* group, a hybrid complex making up the common garden irises. It was probably introduced to the Algarve from southern

Arabia by the Moslems, as it was planted by them in graveyards. Among those with white flowers are found a small number of plants which have blue or blue–white flowers; these may be crosses with other *I.* × *germanica* stock. Further exotics from South Africa, naturalized along the roads, and encouraged by further planting by the JAE, *Junta Autonomia de Estradas*, and found particularly in western parts of the Algarve, are two kinds of *Pelargonium*, one with aromatic foliage, and a 'regal' pelargonium, as well as *Watsonia meriana* (Iridaceae), growing to 2 m high and bearing large clusters of red flowers in tall inflorescences above iris-like leaves.

Perhaps the most striking of the southern African introductions throughout the province is the 'arum-lily', *Zantedeschia aethiopica* (Araceae). Native in the Transvaal, it has been grown in Europe since the early eighteenth century for its white spathes, used in the Algarve, as in northern Europe, as a cut-flower, particularly for the decoration of churches. It is widely planted in Algarve gardens, and it is a common feature of roadsides and ditches, where it attains 1 m in height, flowering until the end of May and often again in October and November. A native plant, also flowering in May along roadsides, particularly shady ones, is the statuesque *Acanthus mollis* (Acanthaceae). The leaves of the related *A. spinosus* from the eastern Mediterranean are said to have been the inspiration for the leaf motif in Corinthian capitals.

More common are sheets of *Gladiolus italicus* (*G. segetum*, Plates 123 and 124), found particularly in cultivated areas. This is a red- to pink-flowered species, distinguishable from other wild gladioli in that the anthers are longer than the filaments. It is not clear in the whole of its range where this plant is native, as it is always found in cultivated areas. In most populations there are forms with aborted anthers and, in the Algarve, plants with corms of 1 cm diameter instead of flowers have been recorded at Cabeço do Estevão, near Alcalar for example. Of species which are possibly native, the most spectacular is the blue *Gynandriris sisyrinchium*, the

Figure 95   *Acanthus mollis,*

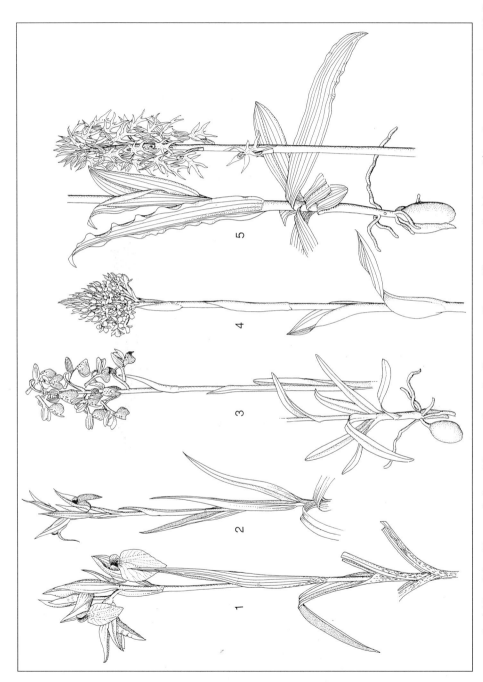

Figure 96 Orchids of the Algarve I. Orquideas. 1. *Serapias lingua*; 2. *S. parviflora*; 3. *Orchis morio*; 4. *Anacamptis pyramidalis*;
5. *Orchis italica*.

'Barbary Nut', one of two European species of a predominantly southern African genus of some nine species. It is more closely allied to *Moraea* than to *Iris*, in which genus it was originally described by Linnaeus. The two northern-hemisphere species are closely related to one another and are considered the most advanced in the genus. *G. sisyrinchium* has flowers opening between 15.00 hours and midnight (earlier in some forms). Most species in the genus have a similar pattern, although *G. hesperantha* and *G. simulans* from southern Africa do not open until 16.30 in some forms, and fade at about 19.00 the same day, whilst the southern African *G. pritzeliana*, opening 11.00–12.00 noon, has faded by 16.00–16.30. *G. sisyrinchium* grows from the seashore and open woodland to desert conditions across its range, but it possibly owes its spread to human intervention for, in many places, as in the Algarve, it is associated with man-made landscapes.

Flowering from October onwards in the Algarve, much earlier than *G. sisyrinchium*, is the not very showy *Iris planifolia* (*I. alata*, *lírio de amor perfeito*, Plate 125), one of the remarkable and beautiful Juno Group of irises, whose main area of distribution lies in Iran, Iraq, and Turkey. In the Algarve it is widely distributed, but only on limy soils and in open fields, often in clusters up to 50 m² in area, rather than by roadsides. The colour of its flower is very variable, ranging from mauve to whitish-blue *I. planifolia* is closely related only to *I. palaestina* — found at low altitudes in Israel as well as along the coast of Lebanon — with which it shares the common feature of spiny pollen grains, these two species forming an isolated taxonomic group.

Plants from the *matos* persisting, even under heavy grazing régimes, include the highly toxic asphodels though sheep seem immune to them. *Asphodelus ramosus* flowers in early spring, and produces fruit capsules 8–15 mm long, followed in summer by *A. aestivus* with capsules less than 8 mm in length. These are the tall species. *A. fistulosus* (Plate 126), found in similar habitats, is a short-lived perennial with hollow leaves, almost round in section. Hybrids between the two tall species are believed to occur, and it has been suggested that *A. ramosus* is itself of hybrid origin, with *A. aestivus* and *A. albus*, a species not recorded in the Algarve, as its parents.

Such conditions, where grazing keeps down aggressive plants, favour the Algarvian orchids of open places, especially species of *Ophrys* and *Orchis*. The pink but variable-coloured and long-lasting spikes of *Orchis italica* (Plate 127) and *O. morio* (Plant 128) are commonly seen, while *Ophrys scolopax* (Plant 129), *O. tenthredinifera* (Plate 130), *O. fusca* (Plate 131), the yellow *O. lutea* (Plates 132 and 133), *O. bombyliflora* (Plate 134), and *O. vernixia* (Plate 135) are plentiful. *Ophrys vernixia*, the 'mirror orchid', is notable for its brilliant metallic blue reflective marking on the lip. It is pollinated by a scoliid wasp *Camposcolia ciliata*, somewhat larger than a honey-bee, which is not attracted in the absence of the lip.

The glistening 'mirror' on the lip resembles the bluish shimmering of the wings of the female wasp at rest. The flowers produce an irresistible scent (mainly cyclic sesquiterpenic alcohols and hydrocarbons) which attracts the wasps even if the flowers are covered. The scent is probably mimetic of the female bee sex pheromone, at least in *O. lutea*; that of *O. vernixia* consists mainly of aliphatic acetates, notably octyl acetate. There are more male wasps than female, and they emerge a month before them and attempt to copulate with the flowers, their behaviour being guided by touch

Figure 97 Orchids of the Algarve II. *Orquideas*. 1. *Ophrys apifera*; 2. *O. lutea*; 3. *O. vernixia*; 4. *O. bombyliflora*; 5. *O. scolopax*.

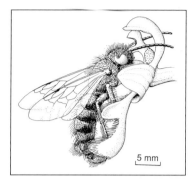

Figure 98   Pseudocopulation of *Camposcolia ciliata* with *Ophrys vernixia.*
(Modified from Proctor and Yeo 1973.)

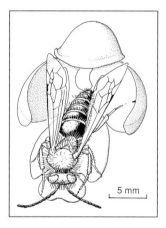

Figure 99   Pseudocopulation of *Andrena* sp. with *Ophrys lutea.* (Modified from
Proctor and Yeo 1973.)

stimuli once they have been attracted by the scent. They deposit sperm and carry off
the pollinia to another flower, thereby effecting pollination.

   Ophrys scolopax, *O. tenthredinifera*, and *O. bombyliflora* are known to be visited
by male *Eucera* and *Tetralonia* bees, also practising pseudocopulation. *O. lutea*, near
Algiers, is also visited by *Andrena* bees but, compared with the wasps on *O. vernixia*,
these alight upside-down for pseudocopulation, the flowers resembling a female
resting upside-down on a yellow flower. If the lip is cut off and reversed, then the
males alight the right way up.

Basketwork

In clearing the *matos* for cultivation, humans sometimes preserve useful plants,
particularly *Chamaerops humilis*. The leaves of this very spiny palm are used in
making baskets and mats. Although increasingly being replaced by cheap plastics,

brooms (*vassouras*) generally flat—are still made with strips of dried palm (*palmita*) for the heads, with the handles traditionally made from oleander wood. Palms for this purpose are cut in May–August, but only during dry weather because the material blackens and is weak if gathered in wet conditions. After drying, the thin strips of palm are soaked in water, following which they are plaited into long strips which are sewn together with *baraçinha*, a fibre from the same palm.

This handicraft basketwork is known as *empreita*, and there is much of it around Alcoutim, Aljezur, Castro Marim, Loulé, Odeleite, Tôr, and São Brás de Alportel. The braided strips are made into a wide range of articles, including *alcofas* (bags for carrying sand etc.), *capachos* (mats), *golpelhas* (the large carrier bags slung across donkeys and mules), *balaios* (hampers), *canastras* (carrier bags, carried on women's heads. and used for salt etc.), *capaz* (for carrying octopus — *polvos*), and *chapéus* ('trilby'-shaped hats worn by local women). Wattle (*vime*), rush (*junco*), and esparto grass (*esparto*), *Stipa tenacissima*, are also used in similar types of basketwork, *esparto* at least having been used for centuries in the Algarve for this. The main places where it was cultivated commercially were near Cape St. Vincent and the Lagos area, for example at Barão de São Miguel; until the time of Dom João III, who reigned from 1521, it was much cultivated around Silves as well and, although it grew only to a relatively short length in these places, it was very durable. A major export market was Spanish Castille, both for raw material and finished work. Another local material used still is fennel (*funcho*, *Foeniculum vulgare*, Plate 136), whose dried stems are made into baskets and mats (*esteiras de funcho*).

Living hedges: succulents

Frequently, when limestone for walls is not available in the *Litoral*, there are living hedges. In the *Sotavento* in particular these are of the 'prickly pear', whose ovoid edible fruits, known locally as *figos do Brasil* (Brazilian figs), are picked in autumn, though care must be taken to avoid the glochids—fine spines—on them. The common Algarve species is *Opuntia maxima* ('*O. ficus-indica*', *figueira da India*, *tuna*, Cactaceae), though the spinier but smaller *O. dillenii* is often preferred as a hedgeplant. Formerly, these cacti were very important commercially in the production of cochineal, a carmine-red natural dye, now largely superseded by synthetic dye-stuffs, an industry started with the discovery in 1856 of the dye mauveine, by the British chemist Perkin. When grown commercially, *Opuntia* is planted at up to 20 000 bushes per ha, and these require regular pruning to maximize production. Apart from dyeing textiles, cochineal is used in the manufacture of cosmetics and foods as well as in drinks, for example in 'Campari'.

Cochineal culture had itself eclipsed kermes oak dye-production — the colorant of which contains kermesic acid. Natural cochineal consists of the dried bodies of scale insects, mainly female since these are more highly coloured and give larger yields than the males, of various species of the Coccidae family, for example *Dactylopius coccus* (*grana*). Indeed Linnaeus referred to these insects as *Coccus cacti*. They feed by sucking the juices from the cacti, and are harvested by scraping them into bags using a long pole with cloth wrapped round it, where they are killed, usually by suffocation, and then boiled in water to extract the dye. A good worker

Figure 100   *Opuntia* hedge near Faro.

can harvest five kg daily of the fresh insects, equivalent to about one-third of that weight when dried, and these contain up to 20 per cent of the dye, carminic acid. Care has to be taken in the harvesting process since the lifecycle of the insects is about 90 days, and it is essential to maintain a stock of the living insects for reproduction. Cochineal has been produced thus, certainly since the nineteenth century, on the Cape Verde Islands, on Madeira, in Mexico, and in Spain. In the Algarve, cochineal (*cochonilha*) was cultivated commercially for many hundreds of years. As late as the 1960s, substantial production of natural cochineal by this laborious means was still being carried on in Teneriffe, Canary Islands, though the major production zone in the world is now around Ayacucho in the Peruvian Andes. Peru's annual production is 220 t of cochineal, all of which is exported, with Japan the largest market, and is worth about US $14 million of a total world market annually for natural and synthetic dyes of about US $200 million.

No species of cactus is native in the Mediterranean, all the familiar species being ultimately derived from the Americas, although the origin of *O. maxima*, which has been cultivated from the earliest times, is still unclear. The more spineless forms of it appear to have been selected for ease of culture and collection of fruit. Spiny forms are more suitable for hedging. Any detached 'pad' from species of *Opuntia* roots readily and this is the general method of propagation.

Although they are completely unrelated, a number of other succulent or spiny plants are referred to as 'cacti' by laymen. Such include a number of lily-allies with long narrow leaves and tall inflorescences like telegraph poles in the genera *Agave* and *Furcraea*, again American genera. Many of these yield a valuable fibre from their leaves, notably sisal from *A. sisalana* of eastern Mexico, which is widely planted in the Old World. Also from Mexico but widely naturalized in the Mediterranean, *Agave americana* is the species commonly seen as a hedge-plant in the *Litoral*. In 1811, it and the cactus, blackberry, and pomegranate were the principal hedge-plants between Albufeira and Portimão; there are vestiges today.

Known as the 'century plant' because of its allegedly long period of adolescence, *A.*

americana flowers after some five or more years in the Algarve. The massive woody inflorescence, up to 6 m high in autumn, heralds the death of the rosette, unless it is cut off rapidly. *A. americana* reproduces from seeds, suckers, offsets, and from bulbils produced in the flowering heads. This change to vegetative reproduction is thought in some parts of the world to be due to frost damage. Such bulbils fall and root, to grow into new plants. The terminal spine of the leaf makes an excellent 'needle' for removing edible snails (*caracóis*) from their shells and, when taken from the leaf complete with attached fibres running out to it, has long been used in the Algarve as a needle-and-thread for rustic repairs. At the end of the eighteenth century, the leaf fibre was an important product.

Figure 101 *Agave americana* hedge near Carrapateira.

Figure 102 Snails at Portimão market.

Also used as hedges and similarly naturalized near the coasts is the related *Furcraea foetida* from tropical America. It also dies after flowering, but in the Algarve the flowers are infertile. It differs from *Agave americana* most obviously in its yellow–green unarmed leaves. *Agave atrovirens*, occasionally naturalized in southern Portugal, has dark-green leaves which are broader than those of *A. americana* and have a terminal spine up to 10 cm long, whilst those of *A. americana* are typically only one-third of this size. *A atrovirens* is naturalized locally in the Algarve, as is the related *Yucca aloifolia*, for example in the dunes behind the beach at Vila Real de Santo António. Like *Furcraea*, it has entire leaves but these are very narrow with a sharp point at the tip. All these plants are grown widely in gardens in the Algarve, as are a number of related species and variegated forms, notably of *A. americana.*

Other roadside hedges

The evergreen *Myoporum tenuifolium* ('*M. acuminatum*', Myoporaceae, Plates 137 and 138), a small tree from Australia and the western Pacific, used in gardens as a hedge, is also planted on roadsides (some good specimens can be seen on the road to Alvôr from Penina). In the *Barlavento*, it is a prominent feature of the local landscape, much of it being seen along the windswept coastal road from Lagos to Sagres, where it grows with the sweetly scented *Pittosporum tobira*, an evergreen from China and Japan. *Myoporum tenuifolium* was probably introduced to Europe at the beginning of the nineteenth century. Elsewhere in the Algarve, the roadsides are planted with Australian species of *Acacia* and *Eucalyptus*, while occasionally are seen specimens of *Bougainvillea*, derived from species mostly introduced from South America in the nineteenth century. Probably naturalized, rather than planted deliberately, however, are quinces, *Cydonia oblonga* (Rosaceae, Plate 139). These fruit trees, *marmeleiros*, are a familiar sight in southern Spain, as well as in the Algarve, with the roadsides in March white with their flowers. The habit in the Algarve, rather than to produce 'quince jelly' in the English way, is to use all the pulp of the large ripe fruits (weighing up to 1 kg each) to make rather hard blocks of *marmelada de marmelos*. It is considered especially valuable for diabetics. *Marmelada* is a leading contender for the much-disputed derivation of the English word 'marmalade'. In the Algarve, the thick paste is used in the same way that jam is spread on bread in England.

Cultivated areas

The cleared cultivated areas have a rich flora of cornfield weeds (Plate 140), which were commonplace in northern Europe 50 years ago but have largely disappeared through intensive farming methods there involving seed-cleaning and treatment, extensive use of herbicides on weeds, and the wholesale grubbing out of hedges. The yellow in March of the *Oxalis* is replaced in April by the even richer yellow of *Chrysanthemum segetum* (Compositae, Plate 141). At field edges too, one sees then the yellow-and-cream *C. coronarium* (Plate 142), often accompanied by red poppies, including the common red poppy of Great War fame, *Papaver rhoeas*, which can colour whole fields, and the blue of *Echium plantagineum* (Boraginaceae, Plate 143),

Figure 103 Plants of cultivated areas (*áreas cultivadas*) I. 1. *Asphodelus ramosus*;
2. *Cynoglossum clandestinum*; 3. *Cachrys trifida*; 4. *Fedia cornucopiae*; 5. *Fumaria
agraria*; 6. *Oxalis pes-caprae*; 7. *Bartsia trixago*; 8. *Scorpiurus vermiculatus*;
9. *Myoporum tenuifolium*; 10. *Acacia retinodes*; 11. *Aristolochia paucinervia*.

Figure 104   Plants of cultivated areas (*áreas cultivadas*) II. 1. *Reseda media*; 2.
Lathyrus ochrus; 3. *Arisarum simorrhinum*; 4. *Ornithopus sativus*;
5. *Chrysanthemum coronarium*; 6. *Lathyrus cicera*; 7. *Nonea vesicaria*; 8. *Echium
plantagineum*; 9. *Cydonia oblonga*.

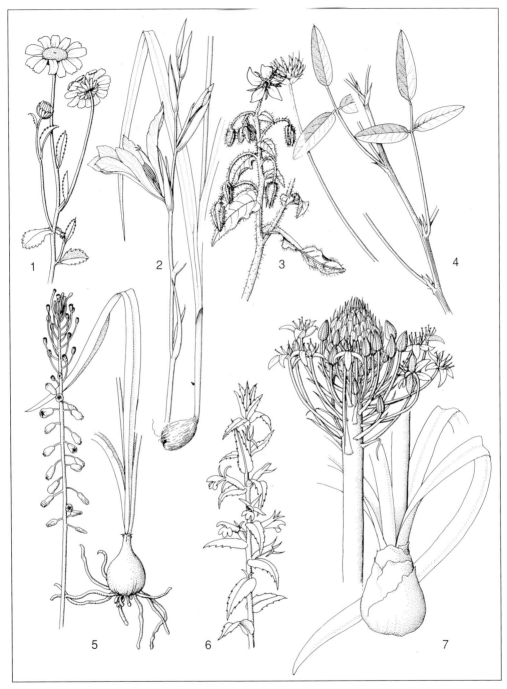

Figure 105   Plants of cultivated areas (*áreas cultivadas*) III. 1. *Chrysanthemum segetum*; 2. *Gladiolus italicus*; 3. *Borago officinalis*; 4. *Bituminaria bituminosa*; 5. *Muscari comosum*; 6. *Bartsia viscosa*; 7. *Scilla peruviana*.

which brightens the fields just as much as the gladioli do. This plant extends into Alentejo, and has, indeed, become a pestilential weed in much of the world. In Australia, it is the ineradicable 'Paterson's Curse' of the eastern states. Another such pest in Australia, and in the *pampas* of South America, is *Silybum marianum* (Compositae, Plate 144), a thistle with white markings on its leaves. Commonly seen in the Algarve, it is an important commercial antidote used in treating poisoning by the toxic fungus *Amanita phalloides*. Similar to *Silybum*, another prickly thistle but with more graceful leaves is *Galactites tomentosa* (Plates 145 and 146). These two are to be found on building-sites and other disturbed ground throughout the Algarve.

A few plants of the *matos* can survive here, notably *Thapsia villosa* and the sand-dune alyssum *Lobularia maritima*, as well as the hemiparasites, *Bartsia trixago* and *Parentucellia viscosa* (Scrophulariaceae, Plate 147). Generally, the suite of species is a Mediterranean ruderal one with legumes such as *Bituminaria bituminosa* (plate 148), *Lathyrus cicera*, *L. clymenum*, *L. ochrus* (Plate 149), *Medicago orbiculata*, *Melilotus indica* (Plate 150), *Scorpiurus muricatus* (Plate 151), *Trifolium angustifolium* (Plate 152), *T. campestre*, *T. stellatum* (Plate 153), and *Vicia* spp. in spring. More striking are a number of *Allium* spp., especially *A. roseum*, *Convolvulus althaeoides* (Plate 154) (*C. tricolor*, Plate 155, flowers later, in May), the snapdragon *Antirrhinum barrelieri*, and 'love-in-a-mist', *Nigella damascena* (Ranunculaceae, Plate 156). Also to be found are *Scandix pecten-veneris* (Umbelliferae), the black-berried nightshade *Solanum nigrum*, flax, *Linum bienne*, spurrey, *Spergularia purpurea* (Caryophyllaceae), and 'quaking grass', *Briza maxima* (Plate 157), *Geropogon hybridus* (Compositae Plate 158) and *Platycapnos spicata* (Fumariaceae), *Aristolochia paucinervis* ('*A. longa*'), and *Cerastium glomeratum*.

Poppies and opium

Of the poppies, the opium poppy *Papaver somniferum* (Plate 159, aptly known in the Algarve as *dormideira*, drowsiness), is commonly seen, wild and cultivated. Under very poor conditions it may be only a few centimetres tall with flowers 1.5 cm across, whereas in a garden the same plant may grow to 1 m in height with flowers 10 cm or more across. It is possible that it is native in the western Mediterranean, or it may be a cultigen derived from *P. setigerum* (*P. somniferum* ssp. *setigerum*) of south-west Asia. It is cultivated, especially from Iran to China, for its alkaloids, which are the basis of a most important drug, opium, unmatched by synthetics in the treatment of intense pain, such as often arises in terminal cancers. The human body is very tolerant of opium, which acts on the central nervous system, over a wide dosage range without damaging side-effects.

Opium is the dried latex, which is obtained from the immature capsules by a light incision made horizontally, and contains over 25 alkaloids including 9–17 per cent morphine and codeine. Opium (*opio*, *anfião*) from the Algarve is said not to have had the same chemical content as that grown in the Levant. It was cultivated commercially for centuries in the Algarve, where forms with white (considered the best for production) and pink flowers were recorded in 1811 as being grown for opium. Yields were greatly reduced by heavy rains in May and June. Much of the opium produced in this way in the Algarve was imported to China by the English, who paid for it by exporting large quantities of tea to England, an early example of 'countertrade'. The seeds, which do not contain opium, are used on bread ('maw-seed', *semente de papolla*), in cake-making, and in some countries as birdseed, the oil from the seeds being used in artists' paints, salad oil, and soap.

Other notable wayside plants

Another most useful herb growing in such places is borage, *Borago officinalis* (Boraginaceae, Plate 160), which is an important plant for bees and formerly was used as a potherb and a flavouring for punch, although in the latter it has now been supplanted by cucumber. A number of other plants in its family are common in waste-places, and these include the brown-flowered *Nonea vesicaria* and *Cynoglossum clandestinum*. A remarkable plant of such sites is the 'squirting cucumber' (*Ecballium elaterium*, Cucurbitaceae, Plate 161); its fruits fall off when ripe, and the fruit-wall contracts such that the seeds within are ejected explosively through a basal hole. The fruit is the source of elaterium, a purgative.

Besides the orchids there are a number of other colourful geophytes of which the most striking is *Scilla peruviana* (Plate 162) with its stout blue heads of star-like flowers and untidy strap-like leaves in early April, *Gladiolus illyricus* and *Muscari comosum* (Plate 163), the 'tassle (-headed) hyacinth', a grape-hyacinth which has a terminal tuft of sterile flowers, of which cv. Plumosum (the 'feather hyacinth') with all its flowers sterile, is widely cultivated in northern Europe. The bulbs are edible when cooked, and *M. comosum* is one of the few native Mediterranean plants whose underground parts can be eaten. Earlier in the season (October–November) and extending into March, the paperwhite narcissus, *Narcisus papyraceus* (Plates 164 and 165, possibly a domesticated form of *N. tazetta*) is prominent in the fields, often in large clumps as at Monte Judeu and at Figueira on the EN125 between Penina and Odiáxere. But perhaps the most splendid show of all is due to the short-lived herb *Fedia cornucopiae* (Valerianaceae), which carpets the ground under figs and almonds in many places in the Algarve with its purple-and-pink flowers in March. With increasing development of the Algarve, such plants, which were probably very rare when the region was forested, are becoming very common indeed.

4

Traditional agriculture

The farmed areas of the region south of the acidic rocks of the Algarve form a mosaic of the traditional agroforestry system of fruit-trees with undercropping, sometimes combined with grazing animals, though much of this land is abandoned or lying derelict pending development and plantation-type agriculture of herbaceous crops, or woody ones, notably citrus.

Animals

Humans, sheep, and goats have hierarchical social structures with a single dominant leader, unlike gazelles for example, so domestication of sheep and goats was obvious and easy. Sheep are grazing ungulates, while goats are browsers. From the fossil record and the present distribution and characteristics of the breeds and their relatives from the wild, the principal ancestor of the domestic sheep (*Ovis aries*) is most likely to be the Urial (Oorial, Asiatic Mouflon, *O. orientalis*), with the original centre of domestication being the Aralo-Caspian steppe and Turkestan. The Moufflon (*O. musimon*) of the Balearics is probably an escape from the early stocks of the seventh millenium BC. Similarly, there is evidence that the main ancestor of the domestic goat (*Capra hircus*) is the Bezoar (*C. aegagrus*) of the mountains of south-west Asia. Goats

Figure 106   Algarve goats.

Figure 107   Algarve sheep.

Figure 108   Grazing cattle and goats.

Figure 109   Horse ploughing. (Photograph Travelpress Europe.)

were kept for meat by the eighth millenium BC, though sheep may have been domesticated earlier.

According to the 1979 Algarve census there were 3100 sheep-farmers with 33 000 sheep (*ovinos*), the largest numbers around Alcoutim, Loulé, Silves, and Tavira. There were 3 800 goat-farmers with 22 000 goats (*caprinos*), mostly in the *concelhos* of Alcoutim, Castro Marim, Loulé, Monchique, Silves, and Tavira. Goats, however, are the traditional animals kept in the Algarve, either, as they have long been, in flocks which browse the open *mato*, or in smaller numbers, often as few as two, as part of the small farm system. Goats raised in this way are usually slaughtered at about 45 kg; the meat is highly prized in local dishes such as *cabrito assado*. The milk is used directly and in cheese-making, notably around the town of Monchique. The skins are used in clothing and for furniture coverings, with the hair sometimes being woven into mohair.

Other major groups of animals in the Algarve include 106 000 pigs (*suínos*) spread among 15 000 farmers, mostly in the *concelhos* of Albufeira, Lagos, Monchique, Silves, and Tavira. In 1979 there were 32 000 cattle (*bovinos*) of which 66 per cent were cows, owned by some 6 000 farmers. This had changed little by 1990 for, in Portugal as a whole, each producer of milk averages only 3.88 cows against a European Community median figure of 17. These statistics show the small scale of the husbandry, and give an idea of the way that animals form part of the traditional Algarvian agricultural system. Of larger animals, horses are becoming much less common than earlier in the century, and mules (*mulas, muares*) and donkeys (*burros*) are also declining in numbers since their rôles are increasingly being taken over by tractors, electric waterpumps, and so on.

Figure 110   Mixed agroforestry of the traditional system, with flowering almond.

Figure 111 Traditional agricultural system, late spring.

Figure 112 Haymaking under tree-crops.

Figure 113 Figs in spring with *Oxalis pes-caprae.*

Figure 114 Traditional produce in Portimão market.

The traditional agricultural system

For many centuries, there has been a large number of farmers with small plots, each worked by a single family. When necessary it has been possible to call on other family members and similar farmers in local villages, for example during harvesting.

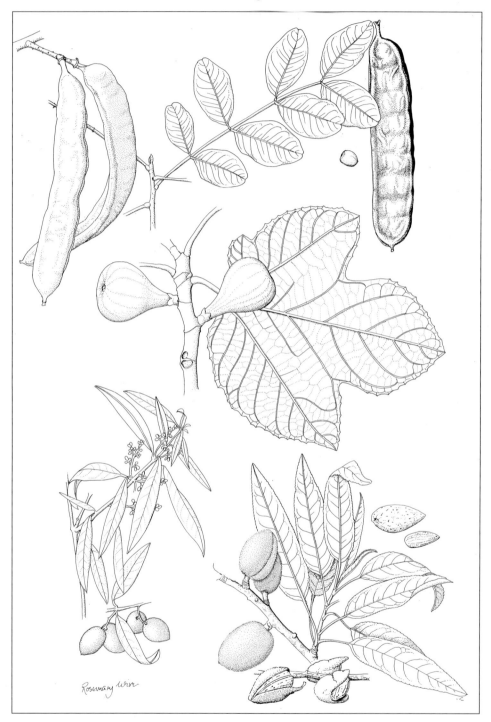

Figure 115 Traditional tree-crops of the Algarve. Top: carob; middle: fig; bottom left: olive; bottom right: almond.

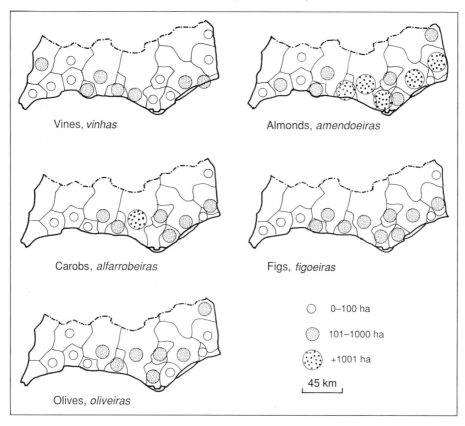

Figure 116 Distributions of almonds, carobs, figs, olives and vines in the Algarve. *Distribuição espacial das áreas de cultivo da vinha e de espécies arbóreas tradicionais do pomar de sequeiro.* (Redrawn from Cavaco 1983, p. 32.)

The Census of 1968 showed that the Algarve then had a total of 34 325 farmers and of these, 20 770, sixty per cent, occupied farms of less than 4 ha, corresponding to 11 per cent of the total area. Ninety-two per cent, 31 770 farmers, occupied farms less than 20 ha, making up 45 per cent of the area, and only 2 555 farmers had a farm larger than 20 ha, with 55 per cent of the farmland between them. Furthermore, the 31 770 farms comprised as many as 134 960 separate properties (*prédios*). Machines were not common, and alongside the animal husbandry, those in the *Litoral* and *Barrocal*, where most of them were, grew fruit trees, usually a 'disordered' combination of almond, carob, fig, and olive, some grapes, possibly a few pomegranates, underplanting the trees with cereals in the winter months, together with field crops such as broad beans and peas. Small animals such as chickens, turkeys, and ducks were allowed to run free near the farmhouse and their manure (*estrume*), with that of the larger animals, was used to fertilize the crops. Water from winter rainfall off the roofs of buildings was stored in *cisternas*. When the farmers were near a hill system, water was drained from there into small earth dams. Local wells (*poços*) were another source of water. If there was any surplus after allowing for domestic needs, this was used to water annual crops grown in a vegetable garden near the farmhouse.

The choice of fruit trees was based on a system which had evolved over several thousand years. The almond–carob–fig–grape–olive assemblage was one which required no water once it had become established; indeed mature trees were often killed if over-watered. Maintenance was easy, since minimal pruning was required, and cultivars that

Figure 117   Typical mats and basketwork of the Algarve.

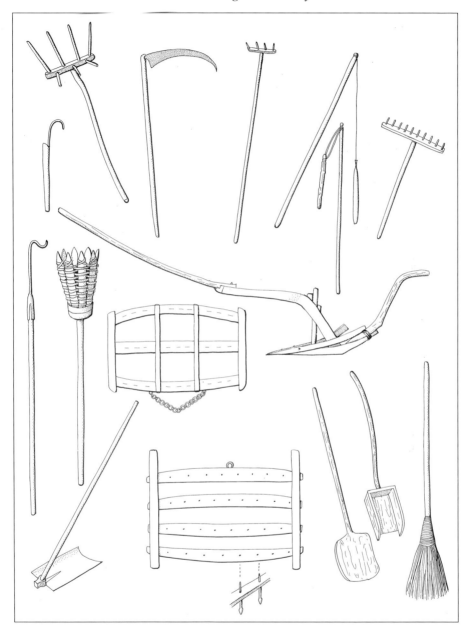

Figure 118 The Algarve: traditional agricultural tools, *alfaias agrícolas*.

were not vulnerable to insect attack were selected. Of these fruits, certainly the dominant one in the traditional system was the fig, closely followed by the vine. Harvesting methods, which made the use of machines unnecessary, had been developed. The making of a wide range of baskets (*cestos*), mats and so on from locally available materials, with which to collect, process, transport, and store the fruits, evolved in association with the growing of those which could be dried or otherwise preserved.

A social system fitted to the agriculture involved cash or barter at co-operative centres both to convert the crops deriving from so many small sources into usable products in a reasonably efficient way and to provide a market. The tree assemblage in the much larger farms in the province north of the Algarve, Alentejo, developed in a different way, consisting mainly of groupings of olives with oaks (*azinheira*, and the cork oak *sobreiro*), although not infrequently with long lines of olives alone, as so often seen in southern Spain.

The unit of measure connected with traditional agriculture is the *arroba* (usually 15 kg but in the range 14.68–15.0 kg). The Spanish *arroba* is 11.5 kg, however, and is actually a *volumetric* measure, equivalent to 16.14 l of wine or 12.56 l of oil. The ancient Algarvian measurements of width are the *braça* (2.20 m) and the *palmo* (0.20 m). Spaniards sometimes still use the *palmar* (0.25 m). The ancient Portuguese foot is the *pé*, equivalent to 0.33 m.

A wide range of tools especially suited to the conditions of the Algarve was developed. These included the giant plough (*arado gigante*), its demise secured by the tractor and so now rarely seen, and the still ubiquitous 'digging-tool', the *enxada*.

To the Algarvian farmer, the moon (*lua*) has always been important. Irrespective of what otherwise would seem to be ideal weather and growing conditions, planting of seeds and of plants as well as all transplanting, are tasks undertaken only on days when there is a waxing moon. The same tradition is still maintained by peasant farmers in other parts of Europe.

Figs (*figos*)

The fig, *Ficus carica* (Moraceae), has been a key part of the food production cycle of the Mediterranean since at least the early Bronze Age (c. 5000 BC), yielding fresh fruit throughout the summer and dried fruit for all-year-round use. This human reliance on figs is not unlike that of other primates in the tropics, where many of the 800 fig species provide their staple food. Most species are evergreen, although *Ficus carica* is deciduous. Typical of the evergreen species is *Ficus elastica*, the familiar 'rubber plant' of Indomalesia, which although perhaps extinct there, is now often to be seen in Algarve gardens. Fig-trees may be monoecious or dioecious, that is to say, with separate male and female flowers on one plant, or with these on separate plants respectively. The flowers are minute and not visible to the casual observer, as they are contained inside the fleshy fig, or 'syconium' (*sycone*), which swells on maturity to resemble a fruit, although technically the fruits are the single-seeded gritty 'pips' inside. The fleshy inflorescence-stalk, or receptacle, has the functions of a more orthodox fruit in attracting dispersal agents, and providing the nutritional reward for them. The well-known laxative effect of 'syrup of figs' may therefore be a mechanism which facilitates the passage of seeds through the gut of the dispersing animal with the minimum of delay and of damage to the seeds.

All *Ficus* species contain a sticky white latex. That of *F. elastica* was formerly an important source of rubber, and that of other species has been used as an effective birdlime. Latex of *F. carica* has been used in the treatment of warts, and since antiquity as a meat tenderizer. In pharmacology it is a substitute for papain, the

Figure 119 Fig-trees with young foliage.　　**Figure 120**　Figs with crazed surface.

digestive enzyme extracted from unripe papaya fruits. It is a skin-irritant, causing dermatitis in some people who are allergic to it.

Pollination

In 340 BC Aristotle knew that young figs would only stay on the tree if insects penetrated them, i.e. if they were pollinated. Each species of fig has a pollinating wasp peculiar to it. The precise mechanism of pollination varies between species, that of *F. carica* having been known since ancient times to be an intricate one. By the sixteenth century, the fundamental rôle of the fig-wasp entering through the eye, the ostiole '*olho-do-figo maduro*', of the fig was recognized and described as '*O milagre da natureza*', the miracle of nature.

Many cultivated figs in southern Europe do not require pollination; the fruits (or at least some of them) develop parthenocarpically — so that the plants can be successfully grown beyond the range of the pollinating insect *Blastophaga psenes* (*Cyneps psenes*). This insect, a Hymenopteran, is black and the female is about 2.5 mm long. In those plants requiring pollination, for example 'Smyrna figs', originally from Turkey, the male 'caprifigs' (*figos de toca*) contain over 200 male flowers, with three to five stamens each, around the apical third of the syconium near the ostiole ('eye') of the fig. In the other two-thirds of the syconium are the short-styled female flowers, in the ovaries of which the young wasps develop. One or more female wasps, possibly attracted to the syconium chemically by pheromones, enter the figs in late spring through the tightly overlapping plates around the ostiole, losing wings, antennae, and perhaps fungal spores and other

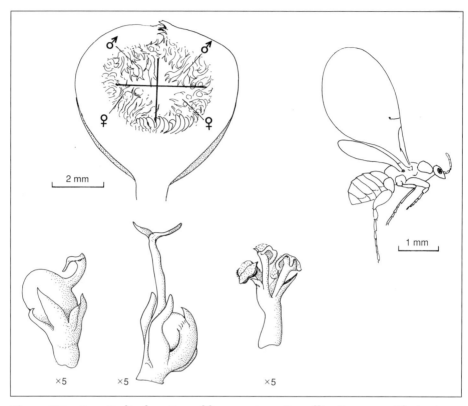

Figure 121   Longitudinal section of fig syconium, its pollinating wasp *Blastophaga psenes* and flowers (left: gall flower; middle: normal female; right: male). (Redrawn from Galil and Neeman 1977.)

pathogens in the process, and then unload the pollen from the figs in which the females were hatched, as a result of swelling and contortion of the body as eggs are laid. The pollen grains are held in intersegmental folds on the insect's body, and many stigmas are pollinated before the wasp dies. In wild figs, the syconia are aborted if pollination fails. The flowers with eggs inside have ovules which are then stimulated to produce tissue on which the hatching larvae feed. The carbon dioxide content in the fig is higher than that outside, and it is believed that this is controlled by bacterial activity, the bacteria themselves being controlled in turn by nematode worms. The relative humidity of the receptive fig is also higher, being almost 100 per cent. The wingless male adults with reduced legs and eyes emerge as the seeds mature, locate the females, copulate, and help the females out of the flowers. Since they are brothers and sisters, this co-operation is understandable in evolutionary terms. Copulation in *F. carica* takes place in normal carbon dioxide conditions as, by then, the ostiole is open, whereas in other species the males are inactivated by the normal atmosphere introduced once they have cut holes in the syconium wall for the females to leave. In *F. carica*, the ostiole opens as the anthers dehisce and block the exit with a mass of pollen. As the female emerges from the ostiole, she cleans off much of the pollen over a period of some fifteen minutes, leaving some in her intersegmental folds. These folds are formed as a result of water-loss with

Figure 122 Fig (cv. Lampeira) with syconium in vertical section.

consequential shrinkage of the insect's body, since the relative humidity within the fig is now well below saturation point. The whole process of fig pollination is repeated by *Blastophaga psenes* in up to three generations a year.

In many species of fig, the system is parasitized by other insects, notably wasps ovipositing through the syconium wall, and therefore not effecting pollination, and by others related to the pollinators coming through the ostiole but not bringing pollen, although they, like many other pollinating species may, unlike *B. psenes* itself, have special but unused pockets for it. There are also weevils and moth larvae, such that in many fig species, up to 80 per cent of the ovules may be killed. They eat developing wasp larvae and mature wasps whilst in the fig. Ants may also feed on the pollinators as they arrive, and some birds eat them as they emerge. Under experimental conditions, *B. psenes* moves freely between 'wild' forms of *F. carica* and trees of closely related non-Mediterranean types, notably *F. palmata* and its allies of north-eastern Africa to India, from which *F. carica* may have been selected.

Although not truly native in the Algarve, figs act as hosts for more lichens than, say, *Eucalyptus*, the bark of which is shed too quickly to allow their establishment. The young fig twigs are often invaded in their first year by a range of species, and crustose ones colonize them when they are about 1.0 cm in diameter. These lichens persist as an extensive cover on much of the bark. Indeed, the stability of the bark ensures that *Parmelia* spp. often attain considerable size, and then the lichen communities can be very species rich, and harbour numerous insects, which often leads to damaged figs.

Cultivars

Many forms of *Ficus carica*, mostly of ancient provenance, are grown in the Algarve. They include the common fig, *figueira comun*, and various cultivars:

'Burjassote'	in both black and white forms
'Cachapo'	very large with rose-coloured skin
'Castelhano'	common in central regions of the Algarve and in the *Sotavento*
'Château Kennedy'	introduced to the Algarve in the twentieth century

'Cotia'	which makes up about 15 per cent of the fig trees in the Algarve, and has yellow–green skin
'Enxario'	a sweet-fruited blue–black fig, which was the cultivar most favoured for drying in the early nineteenth century, though it has to be pollinated; Smyrna figs were not grown in the Algarve at the time
'Eucharia branca'	a white fig of which the black form is 'Eucharia preta'
'Lampeira'	in both white- and now more importantly, black- fruited forms, the latter often grafted on to fig trees of lower quality
'Martinete'	an excellent eating fig
'S. Luiz'	black-fruited
'Urjal'	very large figs
'Verdeal'	a green fig

The Algarvian people use the generic term *figos lampos* for those figs maturing in May–June, whilst those which are gathered in August–September are called *figos vendimos*.

Although cropping of figs is regular — with some cultivars producing two crops each June and August — subject in the Algarve only to rain damaging late-maturing figs, or to crop loss from high winds when they are young, yields can be up to 300 kg per tree. A more typical figure under good conditions is about 50 kg. The weight of individual figs, *lampos* being heaviest, of the best quality (*figo flor, categoría extra*), of which there are only some five per cent in the total harvested, can exceed 125 g. In decreasing order of quality, the remainder of the figs are known as *meia-flor, mercador* (*comadre*), and *industrial* respectively. In general, the dark-fruited and earlier forms, and especially those with dark blue–black skins and dark flesh which mature early (before the end of August), are those chosen for dessert. Later figs, especially those with green skins, are used for drying and further processing into a wide variety of fig-based products.

At least up until the 1920s in the Algarve, fig foliage was fed to livestock, in much the same way that hay (*feno*) is used. The fig harvesting was carried out every eight days or so and, when completed by the end of September, the leaves were gathered. Leaves of *F. semicordata* and other species are still used for cattle-feed in Nepal and elsewhere today.

Propagation and planting

Propagation of figs is possible from seed, and is sometimes practised, but the main method is by grafting selected forms on to seedlings of 'wild' forms. Surprisingly, this is often done on mature trees, for example over 20 years of age, when the trunks have a girth of a metre or more! Grafting is more often carried out when the diameter of the trunk is about 15 cm. The life of a fig tree in the Algarve is about 70–80 years, though in orchards they decline after 50–60 years, but even after dying back, or apparently so, figs generally produce suckers in the following year. Planting density for maximum yield is usually about 64 trees per ha. Figs appear to be widely spaced at such a density, but it allows sufficient space for the horizontal spread of the roots. These are quite different from those of the almonds which have deep vertical tap-roots, as roots of figs spread laterally to at least five times the overhead canopy of the tree.

In western Europe in the nineteenth century, Portuguese figs were considered the best in the world, but figs from Turkey gradually eclipsed them in both European and American markets. In 1850, Portimão, followed by Faro and Lagos, was the principal port exporting dried figs. 1887 records give a total of 25 295 ha of the Algarve planted with figs, yielding 14 000 t. Of these, 4 000 t came from the Lagos region. In 1901, the area was 24 000 ha, i.e. about 1.5 million trees, and they yielded 11 000 t, of which about three-quarters were exported, the main markets being Germany, England, Holland, France, and Belgium. The largest quantity of dried figs exported was 11 895 t in 1922. In the 1940s there were factories at Alcantarilha, Faro, Loulé, and Portimão, exporting specially packaged dried figs for the South American markets. Until the early 1960s, production (from 2.6 million trees) in a good year was still 10 000 t per annum. By 1975, production was down to 2 000 t. It is no coincidence that the *concelho* that has suffered the greatest relative decline and which had the greatest percentage of its area given over to figs in 1901 is that of Albufeira, for it is that zone which has experienced a major expansion of tourism. By 1983, the area devoted to figs in the Algarve had been reduced to 8 000 ha, and the annual production was 1 000 t, with no exports. The downward trend has continued.

Figs on trees are subject to attack by a wide range of pests, most of which may be easily controlled. These include the cochineal insects, *Aspidiotus ficus* and *Ceroplastes rusci* and the aphid *Psylla ficus*. Fungal pests include *Dematophora necatrix*, *Armillariella mellea*, and *Capnodium salicinum*. Since the 1960s, moreover, the fig plantations of the Algarve have been widely neglected, if not grubbed out completely.

Harvesting and processing

Figs, which occur in small clusters throughout the crown of the tree, are harvested by hand-picking. They are mature when they have changed colour — more evident in the

Figure 123   Dried figs.

pear-shaped *lampos* than in the spherical green figs, softened and swollen to full size, and when a slight 'craze-cracking' — not so obvious on the earlier-ripening *lampos* — can be seen on the surface. At the same time, the ostiole opens and the reddening around it largely disappears. When gently lifted and turned through 90°, ripe figs are easily detached from the branches and do not exude any white latex from the stalk. Those left to dry further are processed into 'dried figs' and are knocked off the trees with canes on to plastic sheets. Since ladders are not readily leant against the fragile branches, the very ripe figs were traditionally harvested with a special 'fig-hook' to grasp the twigs, sometimes with a small cane basket held at the top of a short pole into which the fruits were dropped. In the larger orchards, special wooden trestles which did not lean against the branches were used.

Figs for drying are laid on cane mats (sometimes after washing in slightly salty water to which olive oil has been added) and left in the sun, often on flat terraces (*açoteias*) of the older style of Algarvian houses, and turned over daily. They darken appreciably after two weeks or so: the green cultivars turn brown, and the blue ones nearly black. Then, green cultivars of figs are squashed flat, but without rupturing the outer skin, and dried for a few more days. Commercial producers used a similar system, but placed the figs in baskets laid out in layers for drying inside specially constructed *estufas* (greenhouses made from local materials). Figs lose 60–80 per cent of their fresh weight when dried in this way. Formerly, those for export were packed into mats ('tapnets') each containing one *arroba* of 33 lbs, four of them being placed in a '*seron*'. Those intended for products such as 'fig cheese' (*queijo de figo*) are placed in layers in wicker baskets, with cinnamon (*canela*) and aniseed (*erva doce*) sprinkled on successive layers. The mixture is trodden flat every three to four days. Pieces of almond are often added to the layers. The whole mass ferments somewhat and, though the outer skins of the figs stay tough, the inner flesh remains juicy for many months. A wide range of products incorporating fig cheese and broken almond kernels is made in similarly labour-intensive processes. Lower-quality figs are baked in ovens (*fornos*), and then stored in open baskets (*cestas*).

Aguardente de figos (a flavoured alcoholic spirit), is produced by distillation and, for this, aromatic herbs are added to the figs. One *arroba* of good quality figs yields five to six *canadas* (10–12 l) of *aguardente*. *Aguardente fina*, which is absolute alcohol (ethanol), is also produced by distillation though much less is manufactured today than formerly. One hundred kg, of figs yield about 26 l of absolute alcohol, which is used for a wide range of industrial purposes as well as in drinks such as *aguardente de medronho* from the strawberry tree (see Chapter 6), and *absinthe*, which is high-strength alcohol flavoured with wormwood, *Artemisia absinthium* (Compositae). Production of absinthe is banned in many other countries, because of its alleged ill-effects. The residual 'cake' from these processes is compounded into animal feeds, especially for pigs.

Almonds (*amêndoas*)

From the end of December onwards, almond-blossom is one of the magnificent sights of the Algarve. Different cultivars extend the flowering period well into March. The almond, *Prunus dulcis* (Rosaceae), was probably one of the earliest domesticated trees in the Mediterranean. Most cultivars require cross-pollination, although some groups,

Figure 124 Almond blossom.

Figure 125 Almond flowers.

Figure 126 Winter oats under almonds.

Figure 127 Almond cvs. with different flowering times.

Figure 128 Ripe almond fruits.

Figure 129 Almond-harvesting. (Photograph Travelpress Europe.)

Figure 130 Dehusked almonds. (Photograph Travelpress Europe.)

Figure 131 Almonds at different stages of processing.

for example Apulia and Bari almonds, are self-compatible. The edible seed lies within a woody endocarp, like a plum-stone, within the fleshy husk of the fruit, a drupe. Wild forms have high levels of a bitter substance called amygdalin, a glycoside, and a tough endocarp, both affording protection against predators. Through the enzymatic action of emulsin, hydrocyanic acid is released when the cells are damaged through crushing or other injury, and so, in humans, consumption of a few dozen bitter almonds can be fatal. As the non-bitter condition is controlled by a single dominant mutation, this means that some 75 per cent of all seedlings raised from orchards of such trees are also sweet-fruited, and hence growing from seed is the usual method of propagation, though it is sometimes followed by grafting of mature trees.

   The origin of the almond is unclear, because of the number of feral and weedy forms associated with the highly disturbed vegetation of the Mediterranean and Near East, but it seems that its truly wild ancestors are found in the Caucasus. Endocarps have been found in sites from the Palaeolithic onwards. As with the fig, the almond is thus not native in the Algarve, but it too acts as a good host for lichens, which first become established in leaf-, inflorescence-(flower grouping), and bud-scars, forming an almost complete cover. As growth proceeds, the smooth bark is fissured from the new bark underneath it and layers, complete with the lichens, are shed. The bark then bears a different lichen flora. In rain-tracks on older trees, colonies of mosses become established. Large almond trees thus support a varied flora of plants growing on them.

Cultivars

In the Algarve, there are two kinds of almonds, those with hard shells and those with soft shells, known as *durazias* and *côcos* respectively. Formerly, the soft-shelled types were favoured, but this has changed, probably owing to the high-technology machines which now process almonds quickly and cheaply. There are many cultivars, the differences between them being in date and length of the flowering period (which for *côcos* varies between 15 and 39 days), characteristics of the flowers (including the colour), regularity of cropping, and the ease of milling, size, shape, and flavour of the kernels ('nuts', the best weighing up to 2 g), and ratio of kernel weight to that of the shell. The cultivars grown differ in various parts of the province. Around Portimão, for example, 'Bonita' (soft-shelled), 'Convento' (hard-shelled, many with twin kernels, poor quality) and 'Lourencinha' are well-known, though the total number of almond trees in the *concelho* has been sharply reduced, since less than 2 per cent of the region is now planted with almonds–carobs–figs–olives, compared with 10 per cent in 1950. Other almond cultivars in the Algarve include the following:

'Bojuda'	hard-shelled, large kernel, dark skin, good quality
'Ferragudo'	hard-shelled, medium size
'Fôfana'	hard-shelled, pale skin, regular cropper, top quality kernel, many trees around Albufeira
'Grada'	kernel oblong and pale, much grown near Tavira
'José Dias'	large, dark-shelled, grown around Faro
'Ludo'	similar to 'José Dias', grown around Albufeira and Loulé
'Molar de Fuseta'	soft-shelled

With all cultivars, there is considerable annual variation in production. Since 1970 there have been efforts to plant late-flowering cultivars, to reduce problems arising from cold weather, and which have other desirable economic characteristics. The cultivar Ferragnès, developed in France and planted with cv, Ferraduel as a pollinator, is such an important introduction. Late-flowering, self-compatible cultivars recently introduced from Spain include 'Aylés', 'Guara', and 'Moncayo'.

Harvesting and processing

The harvesting method is the same for all cultivars. About the end of August, that is after the fig harvest but before that of the olives, almonds in the Algarve are knocked ôff the trees with long canes (in Spain this system being known as *vareo*). By this time, the sun has generally split open the outer husk, exposing the shell containing the nut, and many more husks break off when the fruits hit the ground. The rest have to be handsorted, or processed through a simple machine. They are then dried in the sun for about two weeks, after which the endocarp comes away from the husk. The drying is traditionally carried out on the flat roofs of the Algarve houses. Then the nuts are bagged and taken to crushing machines, now in ever-decreasing numbers across the province. In such machines, they are fed from a hopper into a set of reciprocating hammers, which break open the shells without crushing the nuts. A series of air-blowers and filters separates the mixture into about ten 'product streams', ranging from whole nuts to powdered fibre. Further machines or hand-sorting isolate valuable products. The uses of the various streams are described below.

Production

The major competition to almond kernels (*miolo*) is from hazelnuts. Even so, almonds remain very important in world commerce and 1987 world production of them, as kernels, was 386 000 t, with a value of about US $4 300 per tonne, the world's largest producers being California 272 000 t and Spain 70 000 t. In the Algarve, the yield for a 30-year-old tree is about 20 kg of almonds in shell (*em casca*); for trees 70 years old, this drops to about 5 kg and to less than half this for still older trees and for those in poor condition. The trees are usually planted at some 64 per hectare, and the traditional mix of plantings in such an arrangement was 20 of *côcos* and 44 *durázias*. Productivity in almond yields, owing to lack of new cultivars and technology in the Algarve, is poor compared with California. There, yields under intensive cultivation in irrigated land can reach 12.5 t per hectare whereas in the arid and underdeveloped conditions of the Algarve, yields are less than one fiftieth of this, at about 0.25 t!

The export of almonds from the Algarve before 1850 was insignificant. In 1898, 5 826 ha in the Algarve was planted with almonds, with a total production of 1 583 t, of which about 10 per cent was consumed locally, the rest being exported, mainly to England, Belgium, Holland, and Germany. Exports between 1935 and 1975 averaged 2000 t (as kernels), the bulk from the eastern Algarve, chiefly the *concelhos* of Olhão, Albufeira—formerly the major area in the province but now largely overtaken by the tourist industry—Castro Marim, Faro, and Alcoutim. In 1965 there were 4.2 million almond trees which gave, as shells, a total of about 10 000 t. Since 1975, production has been falling as the yield of the trees declines with age and lack of maintenance, and there have been few new plantings. By 1976 exports of almonds from the Algarve were only 300 t (in shells), and remained negligible in the 1980s.

Uses

The product streams from the cracking machines have various uses.

1. The most valuable is the whole kernel. The best almonds can yield 1 kg of nuts from 3.2 kg of uncracked shells, but the average ratio in the Algarve today is 1 kg of nuts to 5 kg of shells.
2. The shell fibre is used as a reinforcement in clay tiles.
3. The small broken pieces of nut are used in *pâtisserie* and in various sweets, often with fig and fig-paste.
4. The pieces of shell are used for fuel.

Figure 132
Marzipan and other
almond confections.

5. The nut is used in marzipan, a mixture of powdered almonds with whites of eggs and sugar, and in other confections such as nougats (including the local *torrão de Alicante com amendoim* — much better known in Spain). *Confeites* are sugared almonds, a traditional Easter present.
6. Almond oils are extracted from broken nuts. These are used in perfumery, soaps, and cosmetics, and in medicine provide an excellent solvent for ear-wax.

As grown in the Algarve, almonds require little pruning, other than a trim every ten years or so, to reduce extending laterals, to clear out the centre of the tree, and to remove broken branches. They benefit from both synthetic fertilizers and natural manures of grazing animals for example.

As with figs, added water often kills mature trees. All the forms selected in the Algarve have great resistance to disease and, although the leaves are frequently distorted, the fruits are unaffected, but it is common for resins to be exuded from both the nuts and from fissures in the trunks of the trees, forming hard copals. In the eighteenth century these were used commercially in the province.

Olives (*azeitonas*)

The olive, *Olea europaea* (Oleaceae), unlike the fig and almond, is an evergreen tree, and is the most characteristic and economically one of the most significant of the traditional tree crops in the Mediterranean. Olives provide edible fruits and a 'long-life' oil for cooking, lighting, and medicine, and have been intimately involved for centuries with the prosperity of many Mediterranean peoples. About 30 per cent of the total crop of olives, mostly the less oily ones, are preserved for eating whole, although some cultivars provide fruits which are used for oil as well. The average Portuguese eats 2 kg of olives a year, four times the consumption of the average Frenchman. It is commonly believed that there are two types of olive, 'black' and 'green'. Whilst fruits picked slightly under-ripe are indeed green, those left on the trees for a longer period usually ripen with black streaks, eventually becoming completely black, though some

Figure 133 Olive flowers.

Figure 134 Young olives.

cultivars ripen wine-red, violet, or (rarely) ivory-white. Further processing of the fruits when they are harvested (see below) yields completely black olives. Olives picked green, and processed only in caustic soda and brine solutions, remain green.

The origin of the olive is still obscure, but the rest of the genus *Olea*, some 20 or so species, is largely confined to the Old World tropics, particularly Africa. Indeed, monographers of the family treat the cultivated olive as a subspecies of a wide-ranging tree species extending from Asia to southern Africa. It is suggested that the cultivars have been derived from that African and Asiatic subspecies *cuspidata* (*africana*), which has small drupes and thin flesh. Feral olives ('oleasters') with such fruits occur in the anthropogenic landscape of the Mediterranean, and are fully interfertile with the cultivars, whilst in Saudi Arabia there is at least one form which has extremely sweet flesh. Indeed, Columella (who was born in AD 1) cited sweet olives in Book V of his *De re rústica*. In the Mediterranean today, there is a complex of self-incompatible 'wild forms', oleasters, and clonal cultivars. The olive tree was brought westwards by

Figure 135   Harvesting olives.

Figure 136   Olive graft.

Figure 137   Olive oil crushing.
(Photograph Travelpress Europe.)

the Phoenicians, and in Portugal there is a typical array of cultivated and feral olives. Because of the long time to come into full bearing, which is some 25 years in the Algarve, the felling of olive-trees can have serious consequences, as in the wars between Athens and Sparta at the end of the fifth century BC, when it greatly affected the economies of both states. In the Iberian peninsula, the long life of these trees, 1 000 years being commonplace, has made the handing down of olive trees from father to son a most important part of the cultural and religious heritage. The oldest known olive tree in the Algarve is thought to be that at Pedras d'El-Rei (near Tavira) said to be 2 000 years old. Its girth at 1.3 m above the ground is 10 m.

Although colonization of the twigs by lichens is slower than in almonds and figs, the relatively smooth-barked branches of the olive are colonized by crustose lichens, and bear more macrolichen species than does any other major tree species in the Algarve. Since these lichen colonies provide shelter for many of the insects which attack the olive fruits, special techniques were used to clean the trunks regularly by removing the growths. These included a triangular metal scraper (*raspadeira*) and a large steel glove made from small chain-links (*luva de aço de musgos*).

Harvesting and processing

In the Algarve, the olive grows best on calcareous soils, away from the coast and on land up to 300 m altitude. The olive flowers in May, and has dense clusters of very small yellow–white scented flowers. The flowers are of two types, those with both male and female parts, and those which contain only the male parts. They are wind-pollinated, sometimes self-pollinated as well.

The fruits are ready for harvesting after the figs and the almonds. Unlike figs where the fruit is totally ruined by rain, olives benefit from early autumnal rains as these swell the fruits. Accordingly, it is usually well into October before the crop is gathered. Fruits are harvested by knocking off the olives with long canes or chestnut poles up to 6 m long. This is very laborious since the olives are widely-spaced all over the crown, and individual fruit-weight is low (about 2 g for those crushed for oil but up to 20 g for table olives). When the trees are not on slopes, the fruits are knocked down on to plastic sheets, and gathered into large sacks. These are then sorted, with the better qualities of fruits being processed locally for domestic eating with the bulk taken to co-operatives and crushed for olive oil.

In the former case, after removal of the twigs, leaves, and so on by hand, the olives are treated once in strong brine solution for a few days in unglazed *amphorae* of about 20 l capacity, drained, and then spread out to dry in the sun. This turns the ripe olives completely black. After a further three or four days, they are put in fresh brine solution in the same *amphorae* and the brine solution is changed for fresh liquor daily for a month. The disagreeable solids are leached out of the olive flesh and filter through the walls of the *amphorae*, making the outsides quite black. By then, the olives have a bland taste and are either stored in brine solution to be eaten as a snack before meals, or are further treated by slitting the flesh and immersing them in solutions containing onion, bayleaves, garlic, marjoram, lemon, etc., after which they have a more pungent flavour. Since green olives are picked under-ripe, they contain much more of an alkaloid, oleuropein, than those which are left to ripen on the trees. This bitter substance is leached out with a weak solution of sodium hydroxide before treatment with saline solutions as for the black olives. Finally, all olives are bottled and can be stored for up to five years without significant deterioration if the processing has been carefully carried out using fruits of top quality.

Production

Production of olive oil (*azeite*) is usually in factories, such as the one at Pêra, which handle the olives from a large number of small growers. Such factories are also wholesalers of olive oil and often barterers, trading other traditional products such as figs, almonds, pomegranates, and carobs. The detailed processes used are not the same, but they all include an initial crushing, and subsequent pressing. The crushing is between large stone discs, without cracking the stones, and it produces both a crude olive oil and a mush including the skins known as *bagaço*. The *bagaço* is removed and, in another process, the oil fraction is separated to yield *oleo de bagaço*, used as a cheap cooking oil, with the remaining material distilled to produce *aguardente de bagaço*, a powerful spirit. The crude olive oil fraction is transferred from the stone grinder to a series of hydraulic presses where it is filtered. The olive-stones which have been cracked in the initial crushing release a viscous greyish liquid, *águaruça*, these dregs being left to settle from the oil mixture.

The average oil content by weight of fruit does not exceed 20 per cent. The total yield of oil from the fruits depends not only on the variety, but also on the climate, especially that near harvest time. 100 kg of dried olives of good quality can yield over 45 kg of oil. From the hydraulic pressing comes the top quality 'extra virgin' olive oil (*azeite virgem extra*), of very low acidity — less than 1 per cent. In Portugal, the acidity of olive oils, expressed as oleic acid, is usually recorded on a scale 0, best quality, to 3.3 (*azeite virgem lampante*) which is very acidic, with olive oil mixtures usually marketed at 0.7–1.5 per cent acidity. Owing to disease and neglect of the Algarve trees, there is little of this high quality oil: over 90 per cent of the product exceeds 3 per cent acidity. Further pressings yield a series of oils of ever-increasing acidity, lower quality and value.

The residual pulp from the stones is mixed with water, the low-quality oil floating to the surface being removed, and the leftover solid material is used in animal feeds, or burnt as fuel. This oil is clarified by first mixing with cold water, producing a layer of oil which is skimmed off, or centrifuged, filtered, and bottled. Finally, sometimes it is economic to use solvent extraction of the 'tailings', which gives the lowest quality of oil. In Spain, as well as a higher degree of mechanization of the basic method of olive oil extraction, there is a further process of distillation, and the aroma from this is apparent for tens of kilometres around the factories near Seville for example. This yields a clearer oil than the Portuguese product and also allows the possibility of blending with cheaper oils from other sources, such as sunflower seeds.

Cultivars

Some 96 distinct cultivars of the olive are recognized, and ten of them are commonly grown in Iberia. In Spain, three ('Hojiblanca', 'Lechín de Sevilla', and 'Picual') account for 75 per cent of the olive plantations. The cultivars are largely distinguished by characters of the fruit (including the endocarp) and the leaves. The fruits with the greatest oil content are those of cvs Galega and Redondil.

In the Algarve, the common cultivars are

'Bical de Évora'	vigorous, productive and easy to harvest
'Galega vulgar'	very common in Portugal but with small fruits and not disease-resistant
'Negrinha'	is blackish, the name alludes to the dark leaves
'Zambujeiro'	a regular cropper growing to 12 m tall.

Other cultivars in the province include 'Alcoutim' and 'Longal Brava'. Those used in the production of preserved green olives include

'Cordovil de Castelo Branco'	a strong grower
'Carrasquenha'	late-flowering and late-ripening
'Mançanilha Algarvia'	rounded fruits, common in the Sotavento

When they are grown as a monoculture, olives are planted, sometimes irregularly, about 6–10 m apart with approximately 200 trees per hectare. Propagation is either by seed or grafting, usually by 'crown' (*coroa*), grafts on *Ligustrum* spp, (privets) or on 'wild' forms of olive. All olives benefit greatly from manures, animal and synthetic, ashes, and so on, but those in the Algarve do not require supplementary water. By contrast, all olives in California are irrigated, though they are not grown there for oil production. Cropping is irregular, with a variation of four to one in crop weight between a good and a poor year being typical. When not growing on slopes, olives can be underplanted with crops such as potatoes, which are preferred, or cereals. In recent years, the quality of many of the olives in the Algarve, and hence of the olive oil produced, has been reduced by widespread insect attack of the fruits; losses can exceed 50 per cent of the total crop. Among pests and diseases attacking the Algarve plantations, the olive fruit fly (*a mosca*, *Dacus oleae*), which lays its eggs inside the drupe, where the larvae then hatch out, the olive moth (*a traça*, *Prays oleae*), and autumnal fungal attack from *a gafa* (*Gloeosporium olivarum*) are especially serious.

In 1850, olive culture in the Algarve was said to be neglected, though in 1902 there were still 20 413 ha (8.5 per cent of the then cultivated area of the province), the major production zones in descending order being Tavira, Loulé, Olhão, Faro, Alcoutim, and Monchique. The average annual production of olive oil in Portugal was then more than 82 million l, and it continued at this level to the early 1960s, although with considerable annual variation due to the markedly biennial bearing resulting from the two-year cycle of flower-bud formation. By 1968 the 1.23 million olive trees in the Algarve registered in 1954 had been reduced to 0.96 million and yielded an average of 20 kg of fruit per tree. What seems to have happened since the 1950s is that the olive plantations have progressively degenerated, a major factor being Government control of the price of olive oil, which for a long time was set at levels which gave inadequate returns to the producers, who therefore turned to other crops, or became involved in adulterating the olive oil with oils of lesser quality. By 1979, the area devoted to olives had declined to 2.3 per cent of the cultivated land.

Since the 1960s, the coastal zones, and particularly Tavira, have lost their importance in olive production, as the demands of tourism have removed cultivated land near the sea for projects of more short-term value. Olive production around Silves is now twice that at Tavira, followed by Loulé, Vila Real de Santo António, and Albufeira. Production of olive oil declined from a peak of 14 000 t in 1969 in the *concelho* of Silves to less than 4 000 t by 1979. A similar decline in quantity, and in many cases, a parallel decline in quality, has occurred throughout the ancient olive groves of the province.

By 1947–66, the average annual production of olive oil in Portugal had declined to 40 million l, and was of lower quality than earlier this century. In 1978, the national production of all olive extracts was 41 775 t — about 40 million l, of which 29 789 t was oil (*azeite*), 3 300 t was *óleo de bagaço* (oil extracted from mush) — sometimes sold without further treatment as a cooking oil — and 8 686 t was *óleo de sementes* (oil extracted from kernels). In the same year more than 20 000 t of table olives were sold as well. By 1991, oil production was 26.3 million l, being the yield of some 2.5 million trees.

Uses

Olives represent one of the world's major food crops, with an overall value annually of about US $6 billion. Fine quality olive oil is much used in cooking, particularly for frying and in the preparation of salads, mayonnaise, and also in canning. It finds industrial uses in the manufacture of lubricants, oils, soaps and perfumes, and textiles.

Moreover, Sir Hugh Plat in his *Jewell house of art and nature* (1594), noted a practical method 'to prevent drunkenness — drinke first a good large draught of Sallet Oyle, for that will floate vpon the Wine which you shall drinke, and suppresse the spirites ascending into the braine'! Some modern gastroenterologists are now accepting the long-held view of many people in the Mediterranean region that gut-motility and digestion generally are improved by taking olive oil. Another small-scale medical use of the olive is to prepare infusions made from freshly picked olive leaves and young shoots, which in two to three days is said to reduce hypertension (high blood-pressure). How this works is unclear, though olive oil contains significant amounts of unsaturated fatty acids such as oleic and linoleic acids.

Carobs (*alfarrobas*)

The carob (*alfarrobeira*, *Ceratonia siliqua*, Leguminosae) is another evergreen tree, growing to a height of up to 20 m with a crown diameter of up to 25 m, and pinnate, leathery leaves which minimize transpiration. Like the fig, almond, and olive, the carob is a Mediterranean outlier of an essentially tropical group, this time the caesalpinioid legumes, the two known species of *Ceratonia* being considered relics of a shrunken Indomalesian flora. The second species is the recently described *C. oreothauma* of Oman and Somalia. Flowering in August to October, the carob's reproductive habits, like *Osyris quadripartita* of the *matos*, hark back to its tropical origins. In Portugal, the carob is known as *Pão de São João* and in Germany and

Figure 138 Carob pods and foliage.

Figure 139 Dried carob pods.

Holland as *Johannisbrot*, said to be because St. John ate the fruits in the desert, with the carob possibly being the 'locust' of the Bible. However, there is no mention of the tree in the Old Testament, and the quotations in the New Testament seem to refer to the insect locust.

C. siliqua is grown between latitudes 27°–42° N and 30°–40° S, including much of the coastal zones of the Mediterranean. It is not known as a wild plant, and it is usually dioecious, but occasionally hermaphrodite. The fruits, which are flat leathery pods up to 30 cm in length, each contain 5–18 hard brown seeds, which are 8–11 mm × 6–8 mm × 3–5 mm in size, and develop from flowers in late autumn to mature late in the following summer in the Algarve. The pods (which may be crushed, *triturada*, or ground to a flour, *farinha de grainha*), are sweet-tasting and contain up to 50 per cent sugars, being the basis of a syrup long-used by Arabs for food. They were sold as locust in greengrocers' shops in England as a cheap source of protein — as was whalemeat — after the end of the Second World War. During and after the Spanish Civil War, carob was widely used for making bread. The seeds (*germe de grainha*) are very small by comparison with the pods, comprising 6–11 per cent of the total fruit weight, and were used as jewellers' weights, the original 'carats' (Arabic — *al karrub*).

The carob has been spread by domestic animals in whose faeces seeds germinate. In the Algarve, the plant is widespread on the *Barrocal* but grows from sea-level to the present-day (artificial) tree limit on Foia, 600 m. The bark remains smooth and is shed infrequently as hard scales, jettisoning its lichens with it, so that there is a cyclic succession of them, the sequence being similar to those on figs and almonds.

Harvesting

Since flowering of carobs and the formation of small immature fruits can occur when other fruits are ready for harvesting in autumn, the fruits have to be gathered with care to avoid damage to the younger fruits, and are usually hand-picked as large clusters from ladders set against the trees, rather than by the system used in the Algarve for olives and almonds, of knocking off fruits with long poles. In the Algarve the fruits are dried in the sun, after which they become completely black and can be stored in heaps in the regional carob factories for many months without deterioration. Then the pods are split open by passing them through reciprocating hammers in a machine which has just two product streams, one of crushed husk, the other of the small and much higher value beans, all then bagged for further processing.

The reproductive lifespan of carobs in the Algarve is about 200 years. In 1965 there were 1.6 million carob trees covering 13 000 ha, little changed from the 12 000 ha in 1900, when the recorded production was about 20 000 t per annum. This now averages 30 000 t — about 10 per cent of world production — the extra weight coming from more mature trees and from plantings beginning around the turn of the century.

Uses

A major industry in the province is the extraction from carob seeds of locust bean gum (*farinha de semente de alfarroba*), which can form very viscous solutions. 30 kg of beans yield 3 kg of seeds, and these give 1 kg of gum. It is much used as an additive in foodstuffs such as jellies, jams, and instant puddings. There are various grades, differing from one another in viscosity (in the range 2400 to 4000 centipoises), particle size, hydration rate, and dispersibility. As well as in food, these high-value gums are used in the textile and paper industries, the manufacture of pharmaceuticals and cosmetics, and in mining, water treatment, and oil-drilling. Since the 1980s, Japan has been a large market for carob, where its high protein content and low bacteria count have proved invaluable in babyfoods. The pods, with a high sucrose content, are still exported to England for sale in the health-food market. The colour provides an added bonus when carob is used for blending in 'cooking chocolate' production. In Portugal, carob is also used to impart a 'cola' flavour to drinks and to give a 'chocolate' flavour to ice-creams.

Formerly the hard heartwood was important in England in making high-quality furniture. In antiquity it was used to build temples during the reign of the Egyptian

Pharoah, Rameses III (1190–64 BC), and is good for underwater construction, as in mills. However, the traditional use of carob in the Algarve has been as a supplement to livestock feeds, particularly for pigs, one of which can easily eat 75 kg per annum, as well as for horses and mules, which both have an average annual consumption of 300 kg.

Cultivars

There are rather fewer cultivars of carob than of almond, fig or olive; they are distinguishable by size, colour and shape of fruit, and include

'Alfarroba de Burro'
'Canella'
'Galhosa'
'Mulata'
'Bonita'
'Costella de Vaca'.

The first four were those cultivated in the Algarve in the 1850s, the sugary pulp of 'Mulata' being considered the best-flavoured for eating. All cultivars yield well even when annual rainfall is as low as 300 mm. As it improves pollination, 'Dióica', a cultivar introduced in the 1960s, increases productivity when six trees are planted to about 100 of the older cultivars. A single well-maintained carob tree produces 200–300 kg of fruit, although most trees are not regular croppers. A more typical yield is 50 kg, about 1700 kg per hectare, although an exceptional tree can bear up to 1 t.

The trees in the Algarve require little pruning other than removing dead branches and cleaning out the centres to provide more light. The lepidopteran *Zeuzera pyrina* can be a serious pest of the growing shoots of the carob and the fungus *Oidium ceratoniae*, *oídio*, attacks the leaves, buds, and especially the unripened fruits. Control of both is difficult. In general, supplementary water appears not to be advantageous other than to assist the establishment of young trees. Natural manures, often provided by grazing animals, are very beneficial.

Pomegranates (*romãs*)

Although the pomegranate (*romãzeira*, *Punica granatum*, Punicaceae), has been cultivated since at least the early Bronze Age, it is not much seen in the Algarve and thus cannot be considered as a very important part of the traditional agricultural system, where, in contrast to the almonds, figs, olives, carobs, and grapes, it is usually planted singly or in small numbers.

Nevertheless, it was often planted in the Algarve as a fruiting hedge and in the early nineteenth century it was noted that there were many (also of *Agave americana*, blackberry and *Opuntia*) between Albufeira and Portimão, and some may still be seen by the road from Caldas de Monchique, in Portimão, on the A Rocha Peninsula, and elsewhere.

Figure 140 Pomegranate flowers.

Figure 141 Pomegranate hedge.

The pomegranate is a slow-growing tree up to 4 m tall and is self-pollinated. Like the carob, it is a cultigen, and the only other known species is *P. protopunica* found off southern Arabia (Socotra). The leaves are tipped with extrafloral nectaries exuding the sugars fructose, glucose, and sucrose. The flowers, surrounded by bright green leaves, appear in April–May, and the fruits which can be up to 1 kg in weight, are harvested by hand in autumn. They are ready for picking when the pericarp splits open and exposes the fleshy seeds. The pulp around the seed is eaten. It is easily fermented to grenadine, a cordial, and is sometimes used in *aguardente*, a brandy-like potent spirit. The alkaloids in the bark are considered to be of medicinal value as are the beautiful flowers ('balustine' flowers), whilst the pericarp has been used in tanning in some parts of the Mediterranean. The persistent calyx is thought to have been the inspiration for the crown of King Solomon and thence the crowns of all modern monarchs. Cultivated forms include the dwarf cv. Nana, often grown as a pot-plant in northern Europe.

A pomegranate normally takes ten years to reach full bearing, when it will regularly yield about 20 kg of fruit a year. It can be propagated from seeds, cuttings or suckers. Although some pomegranates are involved in the traditional barter system, and exchanged for other fruits, there are no meaningful statistics for the number grown in the Algarve, largely owing to the small-scale production. There may be as much as 200 t per annum, but there is no likelihood of any significant increase because there are no alternative uses for the fruits and because of the length of time the trees take to reach full bearing. Moreover, the traditional system of which pomegranates form a part is itself disappearing.

Grapes (*uvas*)

The grapevine, *Vitis vinifera* (Vitaceae), is another of the classical fruits of traditional Mediterranean agriculture, providing sugar-rich (15–25 per cent) fruits, easily stored as dried raisins and fermentable into wine. The vine can be grown well beyond the Mediterranean region and was taken by the Romans as far north as Britain, where vineyards have been recently re-established after a long period of decline. In the Mediterranean, vines are grown on a plantation scale as well as part of a traditional homestead small-holding system known in Portuguese as *hortas* (*quintas* when larger).

It is now known that vines have been grown in south-west Spain for almost 4500 years, i.e. since the early Bronze Age. Cultivated grapes appear to be derived from a wild subspecies, 'ssp. *sylvestris*', probably native in the humid forests south of the Caspian Sea and along the southern coast of the Black Sea. The extent of the natural distribution of these wild grapes is now obscured by feral grapes and intercultivar hybrids. Generally, they have smaller acid fruits and rather more globular pips than do the cultivars. They are dioecious, whereas almost all cultivars are self-fertile hermaphrodites, the sex determination being controlled by a single gene. All species of *Vitis*, a northern hemisphere genus with most species in North America, are woody lianes with coiled tendrils in the positions occupied by flower heads in other plants.

Vines produce sterile and fertile branches, the latter growing up to 4 m in one season. They are deciduous, the leaves expanding in March and falling from October to December, by which month the sap has ceased to flow in the Algarve. Most are pruned back to three or four buds on each lateral in January ready for the following season. On the undersurface of the leaves, in the angles between the main veins and laterals, are minute chambers known as domatia, housing tiny mites which keep the leaves clear of fungi and other pests; they are also found on bay-trees, walnuts, coffee, soursop, and many other plants. Flowering begins early in May. The small greenish flowers are pollinated by bees, which take both pollen and nectar. The flesh of the berry is colourless, the skin giving the 'green' or 'black' colour to the grapes, each of which has two seeds in each of its two cells. The fruits of wild grapes are taken by birds, which disperse the seeds.

Figure 142
Vineyard in winter.

Figure 143   Vines
shooting in spring.

Figure 144
Vineyard in spring.

Figure 145   Vine flowers.

Figure 146   Ripe grapes.

Figure 147
Transport of grapes.

Grapes grown in the Algarve are either for table use (*uvas de mesa*), for winemaking (*uvas de vinho*) or, on a very small scale, for drying (*passas de uva*). Most of the eating grapes are consumed fresh by the growers or sold very quickly through the local retail markets. When they have to be kept in good condition, the bunches of grapes are suspended so that the stalks dip into water and can remain in first-class condition for several months without any additives ('*sistema Thomery*').

Figure 148   Red wine of the Algarve.

From 1855 onwards, the grapes of the Algarve were badly affected by mildew (*míldio*, *oídio*). This was followed by severe ravages of *Phylloxera* (vine root-louse, *Daktulosphaira (Viteus) vitifoliae*), an aphid, which was first identified by the Oxford entomologist J. O. Westwood in 1869: it had been introduced from America and was confirmed as being in Portugal in 1871. Rootstocks of American species resistant to attack were used for the cultivars derived from *V. vinifera.* However, many *Vitis* species are interfertile and many cultivars now in production have been derived from crosses involving American species, especially *V. labrusca*, the 'fox' or 'skunk' grape from eastern North America, and can therefore be grown on their own roots. Indeed, many of the vines cultivated in the Algarve are not grafted and some of the best wines are produced from them; the grapes are easily propagated by cuttings. Other cultivars are grafted on to rootstocks of hybrids with *V. berlandieri*, the so-called 'Spanish grape', native in Texas and northern Mexico. *Phylloxera* is no longer a problem for Algarvian vines, but mildews are still major diseases, and spraying up to eight times a year is necessary for satisfactory fruiting.

Production

Algarvian wines were early exported to northern Europe, including Great Britain, and trade expanded at the end of the seventeenth century as hostilities with France increased. The heavy wines of the region survived the voyage better than others but, by 1813, the preference given to Portuguese wines by the Methuen Treaty was abandoned.

In 1891–1903 the Algarve vineyards annually yielded some 2000 l of wine per hectare, the total being over 16 million l per annum. By 1989 the area given over to vines, though declining, was still 3200 ha, mostly in the *Litoral*, with a recorded production of 4. 7 million l, largely red, with some rosé and rather less white. These figures exclude many individual holdings of up to 100 vines, since these are not required to register, and are farmed in the 'traditional' fashion. In 1990, Portuguese production totalled 10.2 million hl from a planted area of 382 000 ha. worked by some 300 000 growers. 80 per cent of the wine from the Algarvian plantations was processed in the four co-operatives, *Adegas Cooperativas*, of Lagôa, Tavira, Portimão, and Lagos. Bringing together over 300 small growers from Loulé to Silves and from Albufeira to the River Arade, the Lagôa co-operative alone accounts for about 50 per cent of the official wine output of the province. Lagos produces another 30 per cent, with Tavira and Portimão about 10 per cent between them. The many privately owned wineries, especially the *Adega Caramujeira* near Porches, make up much of the balance. Marketing is mainly in areas near the co-operatives (about 50 per cent), or in Lisbon (about 30 per cent), with the rest being a minor export, mainly to the United Kingdom, United States, and Belgium.

The intervention of the EC with its 'structural funds' may have a negative effect in that production of what are considered to be 'low-quality' wines in regions such as the Algarve will be discouraged, whilst other areas of Portugal will be favoured, though only if they attain specified standards.

Cultivars

Cultivars grown in the Algarve for table grapes are, in order of maturity, cv. Cardinal, ready in September, having large bunches of very juicy grapes, well-separated from one another, and weighing up to 20 g, with dark blue skins and averaging two pips in each grape; cv. Chasselas Rosé (skin is *cor-de-rosa*, very sweet), cv. Diagalves (black) and cv. Moscatel de Hamburgo (green), all ripe in August; cv. Rosaky, cv. Itália, and cv. Ferral Carpinteiro, all green-skinned and ripening in September;

Cultivars grown in the Algarve for drying to sultanas include cv. Moscatel de Málaga (very large, black) and cv. Sultana (white, without pips). Cv. Dona Maria is also grown for drying, but in the Algarve most are sold as dessert grapes. This cultivar matures mid- to late-September and has large oval-shaped grapes, hanging in loose bunches, the typical weight of a bunch being 600 g with individual grapes weighing 20 g.

Cultivars grown for white wines include cv. Crato Branco and cv. Tamarez, these being picked from the end of August onwards, followed throughout September by the following cultivars grown for red wine: cv. Aragonês (= Roriz; Tempranillo in Spain), cv. Crato Negro, and cv. Trincadeira, cv. Periquita (= Castelão Francês), named after Quinta da Cova da Periquita, where it was first cultivated by José Maria da Fonseca in the 1820s. It is much grown around Tavira and other coastal areas as well as all over Portugal, e.g. near Borba (Alentejo), but especially in Estremadura. It is, perhaps, the leading cultivar in the country as it yields a wine of fair quality acceptable to the non-Portuguese palate. Cv. Negramole has very sweet and juicy grapes, spherical with an average diameter of 1.3 cm, hanging in tight bunches; a bunch weighs about 100 g, individual grapes weighing about 5 g and the juice ferments on the vine, though the fruit does not shrivel. There is some evidence that it is a cross between cv. Pinot Noir and cv. Grenache. It is the most important constituent of 'modern' madeira produced since the phylloxera outbreaks of the 1870s to 1910s and, in the Algarve, it can make up to 60 per cent of some local red wines.

All these cultivars usually have a very high sugar (up to 20 per cent), and thus alcohol, content. They are used to produce 'small wines' for local drinking. Grapes for white wine are crushed and pressed, then fermented at about 20 °C, with bottling beginning in December. Grapes for red wines are fermented with their skins, at about 25 °C, after which they are crushed twice and clarified, with bottling starting in the April following picking. Significant volumes of rosé wines are also made, sometimes by mixing red and white wines or by addition of colouring matter but more correctly by fermenting with grape skins.

Whilst differences in characteristics associated with sensing wine — those of appearance, flavour, and aroma — are generally attributed to genetic differences between vines as well as to differences in the aerial and soil environments in which the vines grow, it may be that the variable character of Algarvian wines can be attributed partly to perturbations of water status at different stages of vine development, since this has recently been shown to affect the concentration of organic acids in harvested grapes. Moreover, the concentration of anthocyanins, which are responsible for the colour in red wines, as well as total soluble phenolics, which give the wine bitterness and astringency, is greater in grapes which have suffered water deficit. Important work using techniques of computerized numerical taxonomy to identify wines, including their method of production, by a number of characteristics, and cultivars of grapes by detailed analysis of their leaf characters, is being carried on in the Portuguese National Centre of Agronomy (*Estação Agronomica Nacional*), Oeiras.

Some fortified wines, such as *Afonso III* and *Algar Seco*, somewhat similar to sherries in production and taste, are made at the Lagoa co-operative. A secondary product is *bagaceira*, a brandy — colourless to chocolate depending on the details of the production process — made by further crushing of wine grapes, followed by distillation which yields yet another form of *aguardente* in the same way that *marc* is made. *Aguardente* is more usually distilled from the fruits of the strawberry tree (see Chapter 6).

Cereals (Gramineae)

The Algarve is not a major cereal-growing region, but winter wheat (*Triticum aestivum*) and oats (*Avena sativa*), especially at the small *quinta* level, are local crops, as is rye (*Secale cereale*) on the poor shaly soils (see Chapter 6) of the *Serra*. Whilst the level is low and declining, cereals have been cultivated by the small farmers (*agricultores*) in the Algarve since the end of the thirteenth century. Winter cereals are especially easy to grow, since they do not require any watering, and in fields of up to 1 ha can be harvested by hand-scything, followed by winnowing. These cereals are then used for baking bread in the characteristic Algarvian stone ovens.

Maize (*milho*)

Although grown under irrigation as early as 1850, maize, *milho*, is a crop not cultivated on a large scale in the Algarve, but is well known in smaller holdings. The seeds are planted in clusters on a grid of 1 m in March, and the plants grow very tall with very strong intertwined rooting systems anchoring them in the ground. Little is eaten fresh as 'corn on the cob' (*milho doce*), that generally grown in the Algarve being ground for bread-making, often with a large pestle and mortar. Much

Figure 149   Rice-fields, near Aljezur.

Figure 150   Rice-fields, near Mexilhoeira Grande.

maize is fed to the pigs which form a major component of the local integrated system.

Rice (*arroz*)

More significant in the landscape, however, are rice-fields, especially in the western coastal belt. Rice, *arroz*, *Oryza sativa*, was probably derived from *O. rufipogon* in Asia, though there were several centres of domestication and it was selected from the weeds growing in taro (*Colocasia esculenta*, Araceae) fields. The many cultivars are grouped as 'lowland', requiring inundation, and 'upland', or 'dry' rice. Some are glutinous and are familiar in rice-pudding, but the non-glutinous possibly comprise the world's most important food-plant, with 2.7 billion people being dependent on rice as a staple: annual world production is about 460 million t.

Pearl rice is dehusked and in the process much of the protein and vitamins is removed. The flour is used in cooking and breakfast foods. It is eaten with curries in southern Asia, and is the traditional basis of *paella* and *risotto* in southern Europe. Local dishes in the Algarve include chicken, *frango*, stewed with rice, and many fish dishes made from the still plentiful catches of the Atlantic. Speciality dishes based on rice include *Arroz de berbigão* (with clams), *Arroz de lingueirão* (made from 'razor' shellfish, *lingueirão*, *Ensis siliqua*), and *Arroz de polvo* (with octopus).

Rice grown in the Algarve is all of the 'lowland' type and its irrigation is intimately linked to the system of *albufeiras*, the small reservoirs trapping water which cascades in the rainy season from the southern flank of the Serra de Monchique and the Serra do Caldeirão, at the foot of the *serras*, and the aqueducts, many now poorly maintained, criss-crossing the province. However, the rice-fields on the western coast, such as those near Aljezur, remain productive, since the inroads made by the tourist industry have not yet had much effect in the *Barlavento*.

In Portugal in 1854, rice production was 5207 t from 3385 ha, and by 1964 this had risen steadily to 181361 t from 37767 ha, of which 1057 ha was in the Algarve. By 1987, the area had been reduced to 32230 ha, and the yield was 144416 t. However, rice cultivars such as IR36 can yield 8–9 t per ha for each of three crops a year. In the Algarve, ricelands now occupy only 371 ha, much of that at *Vale de Dom Sancho* (Aljezur), with its water canal-system connected to the *Barragem de Mira*.

Cucurbits (Cucurbitaceae)

Cucumbers (*pepinos*)

Important too in the local Algarvian diet are cucumbers, *Cucumis sativus*, another Old World plant perhaps wild in the Sino-Himalayan region. Cucumbers were introduced to Europe at the end of the Roman period and Pliny discussed their forcing. Those grown in the Algarve are 'ridge' types and they ripen from June onwards, producing very large crops at ground level. Apart from a pause in the hot summer months, fruiting continues until the end of October. Cultivars grown include 'Ashley',

'Marketer', 'Palomar', and their hybrids. Such cucumbers are the basis of *gaspacho*, a cold soup better known in Spain.

Melons (melões)

Cucumis melo, said to originate in Africa, includes all the cultivated melons arranged in seven interfertile groupings, the names of which are much confused in commerce. In the Algarve most cultivars are eaten fresh, the seeds also being edible when salted, when they are called *pevides* or *salgadas*. They include those with a scaly deeply-grooved rind (the true cantaloupe) with yellow fresh, bright yellow ones with paler and less tasty flesh (in the Inodorus group) and the late-fruiting — in September — smooth but blotched ones with aromatic flesh. These 'musk-melons' keep until February, provided they are sun dried and stored in airy conditions. Besides these are the much smaller thin-green-skinned melons of the Reticulatus group, ready from June onwards known as 'Ogen' — or variants of that name — selected in Israel. In general, melons, especially the late-fruiting cultivars, are unsuitable for production in the Algarve unless they can be shaded to prevent sunscorch.

Pumpkins (abóboras)

Visitors to markets will inevitably see pumpkins (*abóbora*, *Cucurbita* spp.), which are some of the most important members of their family in the traditional Algarve home. The flesh is used as a thickener for soups and much is crystallized. The fruits may exceed 20 kg (up to 100 kg in other countries, and therefore the largest fruits in the world) and are ripe in September, when they are sun baked on walls, flat terraces, and roofs. The flesh does not ferment, even when the seeds germinate *in situ*. The fruit contains a carboxypyrrolidine called cucurbitin, which is used for worming.

Pumpkins formed part of the beans–maize–squash culture of pre-Columbian central America, but the terms pumpkin and squash are used for different cultigens in different places. There are four major types in the Algarve, where they are largely

Figure 151
Pumpkin.
(Photograph
Travelpress Europe.)

grown in *hortas* and not on a field-scale. These include a deeply-ribbed type and another with a rather warty skin, both with ochre flesh.

The related *Cucurbita pepo* ('*Abóbora açucarada do Brasil*') includes the vegetable marrow which grows well although it is rarely seen, and certain other cultivars also called pumpkins, but most of these are non-keeping and therefore of little interest to Algarvians. Widely grown in the province, as well as in Spain, is *C. ficifolia* ('*Abóbora Chila*', '*Gila*'), a pale-green pumpkin with yellow stripes and pale- to chestnut-brown seeds. In Peru seeds from this plant have been dated to 2000 BC. The other species grown are *C. moschata* ('*Abóbora Napolitana*') and *C. maxima* ('*Abóbora amarela grande*'). All pumpkins have flesh with some 90 per cent water, 8 per cent carbohydrate and 1 per cent protein, whilst the seeds contain some 40 per cent oil and 30 per cent protein and are eaten salted and fried.

Water melons (melancias)

Also very common in the Algarve is the water melon, *melancia*, *Citrullus lanatus*. This has been cultivated since the Bronze Age, and was probably selected from the colocynth, *C. colocynthis*, wild in the drier areas of Africa, and used for its purgative fruits since the time of the Assyrians. Water melon is very widely grown in warm parts of the world. Its red-to-pink flesh is a refreshing food-cum-drink in hot weather. China, Ukraine etc., and Turkey are the largest producers with an annual production each exceeding 4 million t. In the Algarve, the common fruit is dark-green-skinned with black seeds, maturing in July and August, and is the one familiar throughout the Mediterranean.

Although grown on a field-scale in Spain — which has about 30 000 ha yielding some 550 000 t annually — and in the Alentejo province of Portugal, water melons are produced on quite a large scale in the Algarve in *hortas* where they are planted at about 10 000 plants per ha. Like most cucurbits in the region, they are sown directly in the ground and are not trained up posts or wires. Three or four fruits are formed on each plant, and water often makes up over 93 per cent by weight, with only five per

Figure 152
Germinating *xu-xu*.

cent carbohydrate and 0.5 per cent protein. Portuguese cultivars include 'Almeirim', 'Da Covilha', 'Nacional', 'Panónia', Rodeo', and 'Sugar Baby', which differ in the size, shape, and colour of their fruits, and the ripening time.

Other cucurbits

The *xu-xu* (Madeira marrow, *chu-chu*, *cho-cho*, *Sechium edule*), originally from Mexico, is a rampant climber with large pear-shaped fruits, each up to 2 kg, and is easily grown in the Algarve. It is seen in the markets from September to Christmas, after which time it stops growing. It is propagated from the single large flat seed sown separately or, better, by planting the whole fruit. Two distinct types are grown in the Algarve, one with a relatively smooth green skin and with pale-green flesh, the other with an uneven and rough dirty-white skin with white flesh, though their taste and size are similar. After peeling, all fruits are cooked by boiling in lightly salted water for a few minutes.

Another ancient cucurbit cultigen much grown in the Algarve, especially for keeping since its skin is hardened by the sun, is the bottle gourd or calabash, *Lagenaria siceraria* ('*Abóbora carneira*'). Cultivars found in the province include 'Cabaça', 'Cabaça Sifão', and 'Cabacinha Chata da Córsega'.

Legumes (Leguminosae)

As part of the ancient agroforestry system of the Algarve, a number of leguminous crops are grown under almond–carob–fig–olive cultures, most notably peas and broad beans (*favas*, *Vicia faba*), which were important in the medieval rotational system of Great Britain for example. Both are ancient cultigens, though it is extraordinary that the pea is not mentioned in the Bible (chick-pea is the plant indicated by 'pea' there). These are the world's two most important pulse crops. The third is the chick-pea itself (*grão*, *Cicer arietum*), often found in Algarvian soups and stews. It was probably selected in the Near East and is the 'salted provender' of Isaiah.

Figure 153 Peas grown under figs.

Figure 154   Broad beans grown under almonds.

Broad beans (favas)

Fossil evidence of broad beans in the Mediterranean region dates from 3000 BC. In the Algarve today they are sown in 'blocks' from late autumn, and grow to 1.5 m tall. The beans are ready from early spring, and generally before the blackfly (an aphid) has caused any serious damage to the crop. Cultivars such as 'Aquadulce' and 'Algarvia' are very disease-resistant. Such plants are taprooted, and require little added water under normal climatic conditions. As well as being used as side vegetables in Algarvian cuisine, *favas* are the basis of a wide variety of dishes, for example, *Favas à Algarvia* — fried with *toucinho* (salted bacon) and *morcela*, black pudding. They are also dried, salted, and eaten like melon seeds.

Peanuts (amendoim)

Another legume, the peanut (monkey-nut, or groundnut, *Arachis hypogaea*), is cultivated on a field-scale near Aljezur and in other small areas in the Algarve. The flowers bend down after fertilization and their stalks elongate so that the fruit develops underground. Besides being eaten fresh or salted, peanuts are a source of cooking oil. The residual cake is good cattlefeed, while the shells are used as insulation material or burnt and returned to the soil. The plant is a complicated hybrid of species native in South America, and it was introduced to Europe early in the nineteenth century.

Other traditionally grown plants

Bay (louro)

The bay tree (*Laurus nobilis*, Lauraceae) is grown all over the province, usually as isolated specimens. Bay is native in the Mediterranean, and is a relic of the old ever-green forests of the region. The trees are slow growing, reaching 20 m in the wild. Propagation is by rooting cuttings. The leaves are much used in stews and are typical of traditional Algarvian cuisine. They formed the original crown of laurel in Greece,

Figure 155   Cardoon.

Figure 156   *Piripíri.*

hence the terms 'resting on your laurels', baccalaureat, Bachelor of Arts, and Poet Laureate.

Cardoon (cardo)

The stately cardoon, *Cynara cardunculus* (Compositae) a relative of the globe artichoke, though well known in southern Spain, is much less grown in the Algarve. It is useful too as a very large ornamental with its huge deeply divided silvery leaves and easily attains 4 m. When grown as a vegetable, its leaves are cut and trimmed and the succulent midribs removed. After peeling, these can be treated like celery and used in soups. The plants require no watering and are perennial, being cut down to the ground once or twice a year. Propagation is by offsets, or seeds from the large thistle-like heads which form after the beautiful intense-blue flowers have faded.

Chilli peppers (piripíri)

Piripíri is a form of *Capsicum annuum* (Solanaceae) derived from tropical America and familiar as chillies. It is invaluable in Algarvian cooking, being a particularly pungent, small-fruited cultivar used mainly in a dressing on roast chicken, *frango assado*. It grows easily from seed producing a small bush. Whole fruiting branches are cut, hung up, and kept for use as required, the fruits remaining in excellent condition for several years.

Okra (quiabo-de-cheiro)

Although uncommon in the Algarve, okra (*Quiabo-de-cheiro*, *Abelmoschus [Hibiscus] esculentus,* Malvaceae), found throughout the tropics, is easily grown, and

Figure 157 Onion growing.

Figure 158 Onions and garlic for sale.

it does well even on limy soils. It is raised from seed, and would be valuable just for the beauty of its red-eyed sulphur-yellow flowers. Best for cooking are the young fruits which are used as a vegetable or in soups, in which they are excellent 'thickeners', or they can be eaten raw.

Onions and garlic (cebolas, alhos)

Both onions (*Allium cepa*, Liliaceae) and garlic (*A. sativum*) are long-established crops of the Algarve; they were known to the Ancient Egyptians. Neither is grown on a field scale, but none the less they are very important on the integrated small farm. Onions are sown in November–December in nursery beds, the seed usually being selected from first-quality planted bulbs which have flowered in the preceding year. Contrasting with practice in northern Europe, the onions are transplanted about 20 cm apart in large blocks rather than in rows, after the top third of the leaves, and two-thirds of the root system have been cut off with a sharp blade. They are well watered, but are planted near the surface in well-prepared, but not firm, soil. They do not grow well in acidic soils (pH less than 5.8).

Onion cultivars grown in the Algarve include

'Roxa da Madeira'	very large, red-flesh
'Branca de Lisboa'	also grown as 'White Lisbon' in northern Europe, medium-sized, white-skinned,
'De Lamego'	yellow-skinned, very large.

In the Algarve, recipes with onion as a major constituent include *Salada mixta* (onions and tomato tossed in olive oil).

Garlic is planted as cloves (*dentes*, teeth), saved from large firm disease-free bulbs of the previous year's crop. Algarvians favour the more productive rose- and violet-coloured forms, again planting in 'block' form, but the bulbs are not transplanted. Garlic is ready for harvesting towards the end of May, onions from the beginning of July. The leaves of both turn yellow, and the bulbs protrude above the surface. Keeping quality is enhanced if they are sunbaked before pulling up, and carefully laid on the ground for further baking for a week or so before being 'roped' and strung up in an airy place, when they will keep for up to a year, though onions stored for longer are subject to 'neck-rots' (*podridão do bolbo*, *Botrytis allii*), and both they and garlic to sprouting. Yields of onions can reach 35 t per hectare whereas garlic averages 6 t per hectare. Modern cultivars of garlic include cv. Germidour and cv. Messindrome.

In the province, the leek, (*alho francês*, *Allium porrum*), grows well from seed all the year round, with peak harvesting in May, but it is not one of the traditional crops, and is produced on only a relatively small scale.

Sunflowers (girassóis)

In the Algarve, the sunflower, (*Helianthus annuus*, Compositae), is grown on a small scale for bee-forage and oil which is extracted from the seeds, and for the seeds themselves.

Figure 159 Growing sweet potatoes.

After salting and frying in olive oil, they are eaten like those of melons. At Quinta do Marco, near Poço Barreto, there is a larger-scale enterprise growing sunflowers, with the oil being exported, on a 25 ha plot with automatic overhead irrigation.

Sweet potatoes (batatas doces)

The sweet potato (*batata doce, batata da Ilha, Ipomoea batatas*, Convolvulaceae) is one of the old crops of the Algarve, although its importance has much diminished in recent decades. Certainly up to the 1940s, sweet potatoes were much grown all over the province on a commercial scale, about 1000 ha in all, especially around Aljezur, Faro, Portimão, and Quarteira.

Large tubers are planted early in spring and, from the numerous shoots which arise, are cut slips when these are about 10 cm long. They are heeled into trenches at about 45°, and well watered. The plants soon grow into a mass of interlocking prostrate shoots, and the roots swell to form the potatoes, some cultivars bearing tubers bright purple when freshly dug, though they dull on drying. Harvesting is in autumn and an individual tuber can easily weigh 2–3 kg, but the plants grow well only on soils which are not limy. When grown in the Algarve, they have superb flavour and texture and, on a plantation scale, yields are up to 16 t per hectare. There are numerous cultivars, which have been divided into four groups with distinctive physical characteristics, and these are typified by

'Raiz de Cana'	whitish tubers and shaped like the 'root' of *canas, Arundo donax*
'Roxa'	red skin and yellow flesh

Figure 160   Giant cabbage.

Figure 161 New Zealand spinach.

Figure 162 New Zealand spinach, flowers.

'Branca'   white skin
'Lírio'   red–brown skin with yellow flesh.

Formerly about 80 per cent of the crop was used for human consumption, the rest for animal-feed. Sweet potatoes are usually eaten as a vegetable, but, in the Algarve, are much used in cake-making. When grown on a large scale, variations of the following rotation were formerly adopted near Aljezur: rice for the first three years, sweet potatoes the next, and wheat in the fifth.

Other vegetables

Besides the familiar aubergines, carrots, French beans (rarely 'runner' beans), peas, peppers, potatoes, radishes, tomatoes, and turnips, there are also some less familiar vegetables. Notable among these is the giant cabbage, *couve tronchuda*, and indispensable ingredient of many Algarve soups, including the ubiquitous *caldo verde*. This cultivar resembles the '*chou*' of France. In the Algarve, it is grown from seeds planted from December onwards. The seedlings are later transplanted and can attain some 6.0 m with a single stem, from which the succulent young leaves are picked as needed and shredded before being used in the soup. Formerly, this cabbage was used as cattle-feed in northern Europe and until recently the stems in the Channel Islands were made into walking-sticks ('Jersey longjacks'), as well as being used for fencing and fuel.

Another unusual crop is the creeping New Zealand Spinach (*tetragónia*, *espinafre da Nova Zelândia*, *Tetragonia tetragonioides*, Aizoaceae) from eastern Asia and the Pacific, which grows much better in dry climates than does common spinach. Yields can exceed 300 kg per hectare. Many other crops familiar in northern Europe are grown on a small scale for local markets or in cottage gardens for home consumption, although culture methods, as with onions discussed above, very often differ markedly from those practised in the north.

5

Citrus and other cash crops

Citrus (*os citrinos*)

Most conspicuous to the visitor to the Algarve are the citrus groves, the delicious scent of their flowers filling the air from early spring. Citrus has been widely grown in the province for centuries, traditionally in the Silves region but now, with improved irrigation, over much of the Algarve.

Figure 163   Citrus groves near Silves.

Many of the plantations date from the early 1970s, following completion of the large reservoirs at Silves, together with the sinking of many local boreholes when suitable pumping technology and electricity supply became easily available. Such plantings added to the decline of the almond and, more so, the fig plantations, since many farmers felt that the short-term economic gains of citrus, rapidly brought into full-bearing, outweighed the commercial return (*valorização*) from the traditional fruits, much of which had been used only for animal-feed anyway, and which posed problems of labour-intensive harvesting.

Irrigation, especially in the summer months, is necessary to avert water-stress on the trees. Local application systems (*sistemas de rega localizada*), such as microjects ('minispray', *mini-aspersão*) and 'drip' *gota-à-gota*, methods have largely replaced the traditional furrow irrigation and wasteful overhead spray methods of the 1960s.

Apart from irrigation, citrus requires a frost-free climate and grows best on flat level ground, preferably slightly elevated land with good air drainage to avoid frost pockets. Areas which are relatively windy are unsuitable so, other than the western *Barlavento* where winds can be troublesome, citrus can successfully be planted over much of the Algarve, the area around Silves being particularly suitable. Citrus fruits poorly in shady conditions and too much wind and rain can lead to significant damage. For example, over parts of the Algarve, the heavy storms in early December 1987 contributed to a substantial crop loss.

Figure 164 Citrus grove near Alcalar in spring, traditional trees in background.

Although the trees will not tolerate salt-spray, groves can be established from sea-level to 100 m, and on soils with acidity ranging from pH 6 to pH 8. Whilst the traditional tree crops (almond–carob–fig–olive) grow well even on very limy soils, this is generally not so with citrus, although some chalk can be tolerated provided that there is good drainage. Rootstocks (see below) are important here, since take-up of mineral nutrients such as boron and phosphorus is markedly affected by the type used. Thus, the mandarin rootstock *Tangerineira* 'Cleopatra' is tolerant of chalky soils, as well as of saline conditions. Drainage is very important and, in establishing citrus, large rocks must be removed from the subsoil. In the Algarve, citrus trees are generally planted 5 m apart, giving about 400 trees per hectare, and not less than 2 m³ of good soil with a cleared depth of at least 50 cm for each tree is preferable for the root system to develop well in its first five years. When young, most citrus have irregular branching but, even without any pruning, they soon develop into dense uniform trees. Indeed, pruning, other than to remove diseased or dead branches and to open out the centre of the tree, is unnecessary.

Flowering of most citrus crops in the Algarve is from the end of March until the beginning of May, and good weather at that time aids successful fruit-set. Wind-free days in the following months are needed to avoid premature drop of the fruits. Insufficient or inadequate watering in June increases the 'June drop' typical of citrus. Fruits are generally borne singly throughout the tree structure, so leading to some difficulty in development of fruits which are shaded, as well as to additional harvesting costs. They are ready to eat from October onwards, in the sequence tangors, tangerines, and oranges, though many oranges — depending on the cultivar — can be left to hang on the trees for up to three months, during which time they become sweeter and, in the Algarve, late cultivars can be left unpicked until August.

The genus *Citrus* (Rutaceae) comprises evergreen trees native in Asia, but they have been cultivated there for so long and readily form hybrids that it is now impossible to reconstruct accurately the origin of many of them, because the original forests where their ancestors grew have long been felled. The first known citrus in the Mediterranean was the citron, or et(h)rog (*cidrão*, '*Etrogue*'), *Citrus medica*, which is associated with the Jewish Feast of the Tabernacle. It was probably introduced from India. Theophrastus, in the fourth century BC, described its cultivation, and a citrus seed dated as early as 1200 BC has been found in Cyprus. The citron is a small spiny tree with a lemon-like fruit and very acid flesh. Its rind is much used in candied peel but, today, it is rarely cultivated in the Algarve. Nevertheless, it is thought to be one of the handful of *Citrus* species involved in the origin of all the commercially important citrus fruits.

The sixteen or so species of *Citrus* are probably native in south and south-east Asia as far south as the Malay Peninsula. Their simple leaves, sometimes with a winged 'petiole', are considered to be equivalent to the apical tips of related plants with pinnate leaves. The reduced leathery leaves and the fleshy thick-skinned berries with the seeds bathed in juice-filled hairs, making up the segments of what is termed an 'hesperidium', are thought to be adaptations to a rather seasonal climate, which makes them suitable to grow in the Mediterranean climates of the world. Citrus is commercially important in the economies of Spain, Sicily, Morocco, Cyprus, and Israel, as well as Portugal. Indeed, citrus makes up the most important fruit industry in warm countries, especially the southern United States (California, Florida, and Texas), but also northern Australia, Brazil, and the West Indies.

The strongly scented white flowers are bisexual and much visited by insects. Nevertheless, many cultivated citrus, especially oranges and lemons, produce seed without pollination which therefore 'fixes' a genetic line as a clone breeding true from seeds (apoximis), as in oranges such as cvs. Washington Navel and Baia. It follows that the traditional small-scale propagation which was common until 1976 in the Algarve, that is growing trees from seeds which generally breed true to type, is scientifically soundly based. Fruit with the desired qualities is picked when ripe and spread out on level ground to ferment and bake in the sun, after which the seeds are extracted and sown.

Citrus species readily cross and hybrids may also be formed with species in other genera: with *Poncirus* (trifoliolate leaves) to give × *Citroncirus*, × *C. webberi* being the citrange, and with *Fortunella* (kumquats) to give × *Citrofortunella*, the limequats and the calomondin orange (× *C. microcarpa*), a familiar house-plant with small fruits held all the winter. Few of these are grown in the Algarve, except as rootstocks for commercially significant citrus crops.

Rootstocks

In 1842 the incidence of foot-rot and gummosis in citrus grown in the Azores led to the practice of budding scions with desirable properties on to rootstocks which were resistant. In mainland Portugal, a common dwarfing rootstock for grafted oranges is *Poncirus trifoliata*, which is resistant to *tristeza* (*Citrivir viatoris*), a viral infection of citrus carried by the aphid vector *Toxoptera citricidus* (*o piolho castanho*), serious in Spain and elsewhere and leading to die-back of whole trees and the wiping out of large plantations of citrus since control is difficult. However, it is probable that over 90 percent of the sweet oranges in the Algarve, particularly those from plantings made before the early 1980s, are grafted on to the sour orange *Citrus aurantium*, thereby usually producing much bigger and more productive trees, whilst remaining at high risk from *tristeza*. There is also a trend now in the province for established trees of sweet orange cultivars, already budded on to sour orange rootstocks, to be top-worked with potentially much more profitable citrus such as the tangor 'Ortanique' (see below). On the other hand, experiments have shown that sweet orange itself is not a very satisfactory rootstock in the Algarve, since it is prone to gummosis.

Since *P. trifoliata* is not very successful under Algarvian conditions, other rootstocks are being used in the province, including the highly successful citranges (see below), cv. Troyer and cv. Carrizo, both of which lead to virus-resistant large and productive trees and, unlike *P. trifoliata*, have good tolerance to lime up to pH 8.

Oranges (laranjas)

The most important of the Algarvian citrus is the sweet orange (*Citrus sinensis*). Oranges, with other products such as cork, olive-oil and wine, have formed a major part of the Algarve's export trade with the countries of northern Europe since at least 1580. *C. sinensis* probably arose through selection from a hybrid complex between the two wild species, *C. maxima* (the pomelo, possibly native in the Malay Peninsula) and *C. reticulata* (the tangerine, probably native in continental south-east Asia). The word 'orange' is said to be derived from the Sanskrit *nagarunga* and it is likely that the plant was domesticated in China. Discussed in an Indian medical treatise of AD 100 or earlier, the sweet orange was much mentioned in western fifteenth-century literature, and it reached Europe over the Genoese trade routes early in that century. Indeed, it seems likely that travellers brought the orange to Portugal shortly after Vasco da Gama rounded Cape Horn and reached India in 1498, and it is certain that they did from South China by the early 1500s, the Portuguese having arrived there in 1518. According to the testament of João Correia in 1524, oranges were already being grown on Terceira in the Azores. Cultivation of sweet oranges (sour ones had been grown even earlier) rapidly became important in Portugal, soon spreading into other Mediterranean countries. The Portuguese made another major contribution to citrus-growing by introducing a superior variety known as the Portugal orange, of which, according to Valmont de Bomare, the first imported tree ('*Aurantium olysiponense*'), from which all the sweet oranges of the time in Europe were derived, was still growing in the garden of the Count St. Laurent in Lisbon in 1764.

Sweet oranges are the most important citrus crop world-wide, being eaten fresh or canned, and the juice is drunk fresh or preserved. The juice from many navel oranges, such as cv. Baia, rapidly separates into layers, and so is not ideal for juice sold commercially, since this requires a homogeneous liquid for which the cv. Valencia

Late, for example, is suitable. None the less, further commercial exploitation of oranges in the Algarve is inhibited by the lack of juicing and canning factories, which could provide valuable outlets for fresh fruits which cannot be marketed otherwise. A juicing factory at Silves with an annual capacity of about 5000 t of juice is said to be nearly complete, though not yet in production (1991). An indication of the potential market can be seen from Brazil, where annual orange-juice output regularly exceeds 800 000 t, over 90 per cent of which is exported. In the Algarve, parts of the fruit are used in high-quality *pâtisserie*, the rind being sugared and dried to form candied peel, for example. As a byproduct of juicing, it also yields essential oils, extracted by pressing and centrifuging and such oils are important in the perfume and soap industries. Other uses of orange involve the insecticidal properties of the rind, and some soaps are manufactured from the seed-oil.

Cultivars

Famous cultivars include cv. Jaffa — little grown in the Algarve, where it is known as cv. O Val or cv. Chamouti — and cv. Valencia Late which have thin skins and small numbers of seeds. Although common in neighbouring Andalucia, blood oranges are hardly ever seen in the Algarve. One of the most valuable oranges in the province is the *Laranja da Baia*, cv. Baia (Bahia), very similar ones elsewhere being known as 'Washington Navel'. Other cultivars maturing later than cv. Baia perhaps have even greater economic importance. 'Baia' is very common in the Algarve, and is one of the navel oranges which have long been known in Portugal (as *'Umbigo'*), and it is possible that the parent cultivar of 'Baia' was taken by the Portuguese to Bahia, in Brazil. Certainly, it was known early in that state, where the 'Washington Navel' orange originated *c.* 1810–20, and was taken thence to Washington around 1870. 'Washington Navel' is one of the world's most important oranges and is much grown in Australia, California, and South Africa. The large Algarve fruits are rather pale and have a thick peel, which facilitates shipping — though little fruit is exported; they have sweet juice and excellent flavour with not too many seeds. They are generally

Figure 165   Orange cv. Baia.

Figure 166   'Baia' oranges.

Figure 167   'Baia' oranges, unripe.

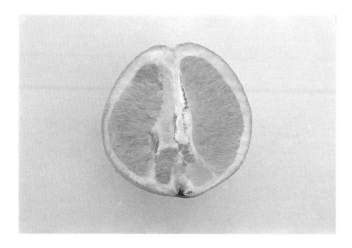

Figure 168   Cross-section of navel orange.

flattened rather than absolutely globose and, where the stalk is attached, there is a deep depression from which shallow grooves run down the surface, but these do not reach the 'navel' at the other end.

'Baia' and 'Baia Rija' are the most important cultivars in the country as a whole, accounting for about a third of the plantings, whilst the following cultivars, with their fruiting times, are widespread in the Algarve and elsewhere in Portugal.

Early

'Newhall', matures from October until January. This is a seedless navel orange with oblong to ellipsoidal fruits of medium-large size and excellent flavour.

'Navelina' (cv. Dalmau), a very vigorous and productive cultivar. It is a seedless spherical juicy navel orange, maturing about two weeks later than cv. Newhall.

Maincrop

'Baia', an excellent orange (see above) which can be harvested between December and March.

Late

'Navelate', a cultivar new to the Algarve, a high-quality navel orange, similar in shape to cv. Baia but the tree grows more vigorously. Fruits can be harvested from the middle of January to the end of May.

'Valencia Late' (cv. Dom João), a widely planted orange of major importance owing to its late maturity; it can be picked from the tree in good condition from April until August. The cultivar enters into production very quickly and gives a high and nearly constant yield of excellent quality rounded fruits.

'Lane Late', recently introduced cultivar from Australia, it is a late-maturing high-quality navel orange ready for harvest from the middle of January until the end of June.

Citrus aurantium, the Seville, sour or bitter orange, *laranjeira azeda*, probably another segregate from the pomelo–tangerine hybrid complex, is less often grown for its own sake in the Algarve. The trees are much more upright than the sweet orange or, indeed, the lemon, and the major branches are armed with strong spines up to 3 cm

Figure 169 Seville orange tree.

Figure 170   Seville oranges.

Figure 171
Marmalade-making.

long. The fruit is shaped like a large tangerine and is harvested from January onwards, when it is still tinged green. It is the basis of marmalade and is blended into liqueurs such as Cointreau and Curaçao as well as *Licor Beirão* produced in the *Serra da Lousan* region of Portugal. The flowers yield 'neroli oil' — *água de azahar*, used in scent. It is mixed with balm oil (*Melissa officinalis*, Labiatae), to give *agua del Carmen*, which is used medicinally as a tonic, antispasmodic, and tranquilizer. The leaves contain phenols which inhibit the growth of weedy species.

Tangerines (tangerinas)

Some oranges have been back-crossed with tangerines to give ortaniques (see below) and other lesser-known fruits, but only a few of these latter are grown in the Algarve. *Citrus reticulata*, the tangerine, named after Tangier, where it was known at an early date, though it did not reach Europe until 1805, can be grown in colder

Figure 172
Tangerine flowers.

climates and has small fruits with readily removable rind and easily separable segments, a feature typical of tangerines and mandarins. Cultivars include clementines and satsumas, which are some of the hardiest of all citrus fruits. Tangerine cultivars in the Algarve are nearly always biennial bearers, whereas oranges, as well as lemons and grapefruit, fruit well annually and with little variation in annual crop weight, provided the trees have not suffered any water-stress and have not been damaged by pests.

Cultivars

Mandarins (manderinas)

'Setúbalense' ('*Citrus deliciosa*', also known as the Mediterranean mandarin), the most important cultivar of this group in the Algarve, with fruits of medium size, many pips, and sweet flesh, very aromatic and maturing between January and March.

Clementines (clementinas, tangerinas de Argélia)

Cultivars are listed in order of ripening.

'Marisol', usually grafted on to rootstocks citrange 'Troyer' and citrange 'Carrizo', it can be harvested from mid-September until the end of October.

'Oroval', a very vigorous tree with large (*c*. 80 g) uniform seedless rounded fruits which are easy to peel, maturing from mid-October until the end of December.

'Clementina' (cv. Clementina Fina), very commonly planted in the Algarve, ripening from November, the medium-sized fruits can be left on the trees without significant deterioration until the end of January.

'Clementina de Nules'; the large seedless fruits are orange-coloured, maturing from November until at least February. It is both productive and early into production.

'Fremont', an early ripener originating from a clementine × 'Ponkan' (a mandarin also known as the Chinese honey orange) cross. The seedless small-medium fruits are brilliantly coloured and have a good aroma, and are ripe from December until March.

'Hernandina' (cv. Hearne); the medium-large fruits, seedy when cross-pollinated, ripen from mid-December until the beginning of March and picking begins when the skins are still tinged green.

Satsumas

These are of little significance in the Algarve, though the following are quite common.

'Okitsu' (cv. Okitsu Wase); originating from Japan, it is early ripening and is harvested from the middle of September until the end of October. It has small, oblate seedless fruits.

'Clausellina', thin skinned and with few seeds, the fruits weighing about 90 g and being quite sweet.

Common hybrids involving tangerines

Those grown in the Algarve include the following.

'Encore', a hybrid tangerine between the tangerine cv. King and the Mediterranean mandarin. Since its harvesting period is between April and August, this permits a valuable extension to the time during which fresh citrus is available in the province

'Fortuna', a tangerine hybrid between clementine cv. Clemantina and clementine cv. Dancy, which ripens between mid-February and mid-May

'Kara', medium to large fruits of good taste and aroma which can be harvested between March and April

'Ortanique', a tangor (see below), the most important in this group

'Wilking', with small fruits, some pips, but very juicy and aromatic, maturing between February and April.

Lemons (limãos)

The lemon, *Citrus limon*, probably derives from a back-cross between the lime and *C. medica*, and is very commonly grown in the Algarve. By AD 903 it was possibly used around Petra near the Dead Sea by the Nabatean people, and was certainly known to the Arabs in the tenth century and introduced to Europe by about AD 1150. It is the toughest of citrus grown in the Algarve and is frequently planted as a shelter-belt around plantations of other citrus. Compared with other citrus crops, its low-branching habit and relatively open crown make the fruit easier to harvest, and pests easier to spray. Lemons have a fruiting habit different from most other citrus fruits in that they flower, and some cultivars bear fruit, all through the year. However, they are mainly harvested in the same overall period as other citrus, December–June.

Cultivars

'Eureka', has smallish fruits, ellipsoidal, very acidic. It is very productive when grafted on to the rootstock 'Macrophylla'. It has a long period of fruiting, from mid-September until the end of May

Figure 173 Lemon tree.

Figure 174 Lemons.

'Lisboa' (cv. Lisbon), with medium-size juicy fruits *c.* 120 g, few seeds, very vigorous and frost-tolerant. In the Algarve, it grows well on *Poncirus trifoliata* and citrange rootstocks and is harvested mainly between October and the end of February

'Lunário', the most commonly grown lemon in the Algarve being very productive. It has large fruits and greenish pulp

'Vila Franca', whilst of little importance in the Algarve is of interest as it fruits throughout the year, hence it is often known as '*Tipo Quatro Estações*' (Four Seasons).

Some lemon cultivars have much more acidic pulp than others. Quite common in the Algarve are cultivars with relatively thick skin (*casca*) and therefore low worth, though valued by Algarvians for tea-making, owing to the essential oils in them. Lemons from a given tree often show greater variation than those fruits of other citrus cultivars. In storage, features of the rind of fruits picked immature are modified more than in other citrus, so that it is difficult to characterize and identify lemons from their fruits alone.

Limes (limas)

The lime was first recorded in the west by Abd-Allatif in the thirteenth century, and in China, much earlier, possibly originating from north-eastern India, adjoining parts of Myanmar (Burma), or northern Malaysia. The lime tree, *limeira*, *C. aurantiifolia*, which produces yellow fruits smaller and rounder than lemons, ready to use when the skins are green, is not now very commonly grown in the Algarve, although it thrives in private gardens. It probably arose from a cross between *C. medica* and another species. It fruits mainly in the spring months, like most citrus in the Algarve. Limes

Figure 175   Limes (left) and lemons.

are rich in ascorbic acid, vitamin C, and were formerly issued to sailors in the Royal Navy to prevent scurvy, hence their nickname 'limey'.

Cultivars

Acid types include cv. Pérsica (the Persian lime *Limeira da Pérsia* or the Tahiti lime *Limeira do Tahiti*), which has medium-large nearly seedless fruits, and is the most common cultivar in the Algarve, being much better adapted to colder climates than 'Mexicana' — the 'West Indian' or 'Key lime' of Florida — with its small round fruits.

Sweet types includes cv. Limão doce da Palestina ('Indian', 'Palestine' lime), which has medium-sized fruits and is the *mitha nimbu* of India, the *succari* of Egypt.

Grapefruit (toranjas)

The grapefruit, *toranjeira*, *Citrus × paradisi*, is not often grown in the Algarve, although plantings increased in the 1960s and early 1970s. It forms a tree smaller than the orange and tangerine. Many of those grown until the 1970s had very thick skins, the pomelo or the shaddock being in their ancestry, but these have largely been replaced by cultivars with thinner skins, sweeter flesh and few seeds and, in some cases, with pink flesh. The grapefruit arose in the West Indies in the eighteenth century, and was probably the result of a back-cross between the pomelo and a plant which itself was a cross between the pomelo and the sweet orange.

Cultivars

'Marsh' (cv. Marsh's Seedless), with white flesh.

'Star Ruby', with pink flesh (*variedade pigmentada rosada*), which has the distinction of being the first citrus cultivar to be patented in the United States.

In the Algarve, grapefruit is harvested between November and April.

Figure 176
Grapefruit tree.

Figure 177
Grapefruit.

Tangors (tângeras)

Less important citrus crops in the Algarve include the tangor, *C. × nobilis*, a sweet orange back-crossed with *C. reticulata*. Intermediate in size between those of its parents, its small round fruit is picked half-green, when it is ready to eat.

To be seen are plantings of cvs. Carvalhal and Ortanique. Cv. Carvalhal fruits from end-September to end-October, two months before the tangerine; tangerines and grapefruit, as well as most cultivars of orange, are much too acidic to be eaten at this stage. Cv. Ortanique is of recent introduction to the Algarve, though it has been grown for some years in Cyprus, Israel, and Jamaica. It is harvested from the end of February until the end of April, so extending the season of the similarly fruited tangerines grown in the province.

Total production of tangors in the Algarve is about 12 000 t per annum.

The tangelo, *C.* × *tangelo*, another minor fruit, arose as a cross between the grapefruit and *C. reticulata*, several such cultivars being widely but incorrectly known as uglis.

'Minneola' (cv. Nova), a hybrid tangelo of grapefruit 'Duncan' and tangerine 'Dancy', has round juicy fruits maturing between mid-December and the end of February when grown on citrange rootstocks 'Troyer' or 'Carrizo'.

Production of citrus

Typical crop-size in the Algarve for a well-grown orange, lemon or tangerine tree over five years old is 50 kg of fruit per annum. Provided an adequate watering, manuring, pruning, and spraying regime is adopted, such trees will continue to bear well for about twenty years, after which production declines. About 49 percent of the oranges in the Algarve are 10–24 years old, and a further 23 per cent are 25–39 years old. Only 10 per cent are over 40 years of age. Yields in the Algarve, admittedly under quite different conditions, are much lower than in the São Paulo state of Brazil, the world's major producing area, where the existing stand of about 110 million mature trees yields more than 9000 million kg of fruit per annum, that is to say, an average of over 80 kg of fruit per tree. The production cost (1990 figures) in São Paulo averages US$1.50 per box of 40.8 kg, that is, less than 4c per kilogramme, with a wholesale price of double this. This degree of efficiency rivals that in the world's other major citrus-producing area, Florida in the United States, with both Brazil and the United States, not to mention Spain, being far ahead of the productivity of the much smaller Algarvian producers, their fruit having a local wholesale price of about US30c per kilogramme! However, production of citrus based on virus-free cultivars on citrange rootstocks on good Algarvian soils, with irrigation using good-quality water can average 50 t per hectare and, with good management, reach 70 t per hectare.

In 1954, there were 3.46 million citrus trees in Portugal, of which only 0.34 million were in the Algarve. At an average spacing of 400 trees per hectare the total area under citrus was then about 8500 ha. Census figures from 1979 show that the Algarve then had a total of 7815 ha of citrus plantations, mostly around Silves (1768 ha), Loulé (1413 ha), Tavira (1402 ha), Faro (1077 ha), Olhão (617 ha), and Albufeira (341 ha). This is about 4 per cent of the agricultural area of the province. In 1987, the recorded production of citrus in mainland Portugal was 105 000 t oranges from 20 501 ha, 18 343 t of lemons from a further 4780 ha, and 16 665 t of tangerines from another 5114 ha, plus 7123 t tangors from 1545 ha. Since the area in 1977 was said to be 25 465 ha, with a recorded production of 153 589 t, it suggests that by 1990 citrus production in Portugal was at least 428 000 t. That of oranges from 9000 ha in the Algarve alone exceeds 150 000 t.

European Community grants are being provided (1992) to establish new plantations where potential production exceeds 20 t per hectare, anything less than this being considered not to be viable. At the same time, regulations have been introduced so that fruit is graded by quality and size. However, such systems are not consonant with the small scale of many of the citrus plantations in the Algarve and the local production and marketing methods.

The most productive area in the province is still the *concelho* of Silves where, of the cultivated area of 2850 ha, 2300 ha is planted with citrus. Of this (1990), 1650 ha is oranges and yields 26 500 t annually; 550 ha is tangors and tangerines (including clementines), yielding 7000 t, and 100 ha of lemons yield 1200 t.

Pests of citrus

Although citrus trees in plantation are usually kept clear of epiphytes, they can, when neglected, support a rich epiphytic flora. All citrus in the Algarve need a comprehensive spraying programme against a wide range of aphids, scale insects, and so on. A serious pest, *mosca branca dos citrinos* (*Aleurothrixus floccosus*), first invaded the groves in 1977, and is still very troublesome, although biological control may remove it. It is a yellow-bodied homopteran parasite *c.* 1.5 mm long, which has four or five generations a year. It mainly attacks the undersides of the young leaves, on which the larvae feed. Whilst this alone weakens the trees and reduces fruit quality and quantity, further damage is caused both by the exudation (*melaço*) from the insects, as well as the dead insects themselves, which later blacken on the leaves and fruits, and lead to an invasion, and more damage, by ants as well as fungi. These ants (*Formiga argentina*) are often associated with aphids which they milk. Aphids (*Aphis citricola* (*A. spiraecola*), *piolho verde da laranjeira*), attack the growing leaves in May and June, and sometimes later as well, leading to rolling and puckering of leaves. Young trees are especially affected by aphids, though mature trees are not immune. Aphids damage some trees but leave parts of the plantation untouched. Infestation can be dramatic and serious.

Just as difficult to eliminate as *mosca branca* is the 'cottony-cushion' scale insect *Icerya purchasi*, a major pest in the Algarve. The damage to citrus leaves and fruits, which yellow, looks much the same as that of *mosca branca*. Many beetles, especially *Vedalia* sp. and the ladybird (*joaninha, Novius cardinalis*), are predators of this parasite and aphids on citrus, but are usually too scarce to effect control. *Pseudococcus citri* (*algodão*) is a very common cochineal-type insect causing similar damage in the province. As in other major citrus-growing areas, the most serious pest of citrus is the Mediterranean fruit-fly, *Ceratitis capitata*. It attacks and ruins the nearly ripe fruits and is active at temperatures over *c.* 16°C; it is a grave problem in the summer when the oranges of cv. Valencia Late are hanging on the trees. From late April spraying once a fortnight is required to maintain any control at all.

The fungus *Colletotrichum gloeosporioides* is responsible for the disease anthracnose, which is widespread on citrus in the province. The fruits are disfigured by black, indented, spots and the tips of the young growing shoots suffer die-back. Field rats (*ratos dos pomares, Pitymis lusitanicus*) and, more importantly, voles (*rato toupeira, Microtus (Pitymis) duodecimcostatus*), attack the underground roots and often kill the trees. Fieldmice chew the bark close to ground level and rabbits also attack the bark and the cambium layer. Poisoning or trapping are used, but neither is very effective. All these pests thrive more in weed-infested ground, where there is other food such as bulbs of *Oxalis*. Citrus is also attacked by snails, slugs, mildew, gummosis, and other bacterial infections as well as viruses and nematodes (small worms), and is sensitive to mineral deficiencies of the soil.

Other fruits

Apples (maçãs)

Apples are ancient cultigens of hybrid ancestry and although always available in the markets, are not grown on a large scale in the Algarve. Nevertheless, fruiting is reasonably satisfactory, especially when the trees are grown away from the coastal zone. Conditions are most favourable around Monchique, with its cooler summer temperature, colder winters and more acidic soil. The most common cultivar is 'Golden Delicious', fruiting in September, with 'Red Delicious' fruiting in November–December, though other cultivars which may be commercially viable have been introduced. These trees are usually grafted on to rootstocks of 'MM 111'. Another apple, hardly grown in the Algarve but usually seen in the markets, though common around Alcobaça in the north, is the superb khaki-skinned cooking apple, cv. Reinéta do Canadá, which is commonly seen as *maça assada*, baked apple.

Figure 178   Apple cv. Reinéta do Canadá.

Other than the cultivars mentioned above, the following are grown in the Algarve.

Apples with red skins (coloração vermelho)
'Fortuna Delicious', 'Royal Red', 'Starkrimson', and 'Top Red'.

Apples with some red (mista vermelho)
'Akane', 'Jersey Mac', 'Reinéta vermelha do Canadá', and 'Stayman Winesap'.

Other apples (outras variedades)
'Grandes dos Boskoop', 'Kidd's Orange Red', and 'Reinéta do Canadá'.

Apricots (alperches, damascos)

The apricot, *Prunus armeniaca*, has been cultivated in China for almost 4000 years and was brought to the Mediterranean in classical times. The fruits are eaten fresh, crystallized, canned or as jam; the juice is sold as a 'nectar', but is also the basis for certain types of brandy and liqueurs. The seed oil is used in cosmetics. The trees flower in early spring and the fruits develop very rapidly, so that apricots are seen soon after the loquats and before the earliest of the peaches. They grow quite successfully in limy soils.

Cultivars

Apricot cultivars are usually grafted on to seedling rootstocks of peach or, more successfully in the Algarve, plum. The fruits mature over a much shorter period than peaches but, like them, cannot be left on the trees.

'Búlida',	sweet fruits,
'Luizet',	very large fruits,
'Maillote Jaune',	golden perfumed pulp,
'Reale d'Imola',	large, red pulp, maturing in June,
'Royal',	large yellow fruits.

Avocado pears (abacates)

The avocado pear (*abacateiro, Persea americana*, Lauraceae) is one of the oldest of all cultivated plants, having been grown for 10 000 years in tropical America, where it is native, and many hundreds of cultivars have been selected. 'Avocado' is derived from its Aztec name *ahuacatl*. They generally grow well in the areas where oranges thrive and, on a commercial scale, in the Algarve near São Bartolomeu de Messines. Since the early 1970s, cultivars have been introduced from the United States of America to the province, and a significant industry has now developed. Small numbers of trees are also to be found in the *quintas* and *hortas*. Total plantings are about 40–50 ha. The climate is not as favourable in the Algarve as it is in Madeira, where the avocado, is more widely grown, thriving there with less variation in temperature, more humidity, and no possibility of frost. Avocados are readily raised from seed, the usual

Figure 179
Avocado pear
flowers.

Figure 180   Avocado cv. Fuerte fruit.

method on Madeira, but trees with desirable characteristics are propagated by grafting, usually on Mexican rootstocks. By careful selection of trees and growing conditions, yields of up to 70 t per hectare have been reported in Israel, though yields in the Algarve for well-managed plantations are about 12–20 t per hectare. To achieve this it is necessary to provide supplementary water in the summer months averaging 1000 l per week for each tree.

The fruit, harvested by clipping, is most familiar sliced in salads or eaten with a dressing of vinegar, oil, salt, and pepper, but it can also be used as a dessert fruit with sugar and the pulp is made into jams. It is one of the most nutritious of all plant-foods, being very rich in oils (up to 20 percent) and vitamins A, B, and E.

There are three distinctive forms of the avocado, though many intermediates are known: one of these intermediates, cv. Fuerte — between Mexican and Guatemalan — is common in the Algarve. Cv. Fuerte normally takes six to seven years to produce significant quantities of fruit.

Mexican

These may be recognised from the leaves which, when crushed, have a characteristic anise scent. They are the hardiest avocados, flowering from December to April, the small fruits with large seeds and distinctively shaped stalks maturing five to six months later. Some cultivars have red or black, rather than the usual green, skins. A common cultivar in southern Spain and the Canaries is 'Topa-Topa'. This cultivar is little seen in the Algarve where however, cv. Bacon is well established; it takes about four years to come into commercial production and is often planted as it is vigorous and somewhat frost-resistant.

Guatemalan

These come from the highlands of Central America and have a stalk which is cylindro-conical at its junction. Cultivars 'Hass', 'Rincón', and 'Taft' are much grown in southern Spain and the Canaries but, of these, only cv. Hass is common in the Algarve. It is self-pollinating and a good cropper with the large nearly spherical fruits, maturing 14 to 17 months after pollination, weighing up to 1 kg, being characterized by thick and woody skins. Cv. Reed ('Anaheim' × 'Nabal') is widely and successfully grown in the province where it was introduced in 1973. Its fruits mature later than cv. Hass — July to September — and it is a heavier cropper.

West Indian

These are widely grown in Florida and comprise the least hardy of the three groups and so are little seen in the province. The fruits are large, generally similar to the Guatemalan, though ripening six to nine months after fruit-set. Most have excellent flavour and leathery skins. Cv. Pollock is a cultivar which is grown in southern Spain but little in the Algarve.

Pests and diseases of avocados

In the avocado-growing areas of the United States and the Canaries, the trees are attacked by a range of pests, including root-rots caused by *Phytophthora cinnamomi* and *Armillariella mellea* (honey fungus) and, most seriously, by thrips, a root louse (*pulgão verde, Heliothrips haemorrhoidalis*), as well as by cochineal insects and aphids. 'Sun-blotch' (*escaldão*), originating from a virus, leads to stunting and reduced yields. In the Algarve, most of these problems are minimized by good husbandry and ensuring good soil drainage; damage by air-frost is perhaps the biggest problem in the province.

Kaki fruit (dióspiros)

The kaki, Japanese date plum or persimmon (*Diospyros kaki*, Ebenaceae), much grown in Japan and China for its edible fruit, and often dried there as well as being an important sugar-source, is planted on a small scale in the Algarve where it is usually grafted on *Diospyros lotus* rootstocks. The fruits are astringent until almost 'squelchy' ripe in a short season in October.

Cultivars

'Berti', synonymous with cv. Coroa de Rei; it has very large fruits
'Faufau', has fruits which are not astringent and is common in the Algarve
'Kouroukouma', has smooth-skinned large fruits.

Loquats (nêsperas)

A more common, elegant fruit-tree grown in small plantations or as a single tree in *hortas* is the Japanese medlar or loquat (*nêspereira de Japão, Eriobotrya japonica*,

Figure 181   Loquat tree.

Figure 182   Loquat fruits.

Rosaceae). It was introduced in the late eighteenth century from China. It is an evergreen, with leaves similar to those of *Rhododendron*, and is sometimes cultivated for ornament in northern Europe. Its dirty-white to pale-yellow fragrant flowers appear in the Algarve in late autumn, the flowering season lasting several weeks.

The yellow fruits, maturing at the end of April to mid-May, are eaten fresh, stewed or in preserves. If they are picked only slightly under-ripe, they are very acidic. In the Algarve, the fruit is very susceptible to fungal scab, *pedrado*, associated with *Fusicladium eriobotryae*. Usually, the symptoms of this disease are restricted to unsightly blemishes on the skins of the fruits. Here it is rarely treated, although in Alicante and Andalucia, the problem has been dealt with by selection of cultivars resistant to the disease, and these also have much larger fruits (up to 60 g). In Spain, where the area under loquats is about 2 000 ha with an annual production exceeding 25 000 t, the rootstock in general use is *Cydonia oblonga*, the common quince; cultivars often seen are cv. Algerie, the cultivar on which most plantations are based, cv. Golden Nuger, cv. Magdall, cv. Nadal, cv. Peluce, and cv. Polop.

Passion fruit (maracujá)

Passion fruit or purple granadilla (*Passiflora edulis*, Passifloraceae), is grown on a small scale in the Algarve. It is a native of southern Brazil and was widely distributed

Figure 183 Passion fruit flower.

Figure 184 Passion fruit, unripe.

throughout the tropics and sub-tropics during the nineteenth century. The climbing plants are grown from seed, and its large golf-ball-sized fruits follow the beautiful flowers which fade rapidly. Flowering begins in early spring and, by June, the ripe hard-skinned and cratered fruits drop off. The sloppy orange-coloured fruit pulp is astringent and valuable for juice, sherbets, ice-cream, and so on.

Cultivars

'Maracujá do Brasil', has dark fruits and is very juicy
'Maracujá da Madeira', has yellow fruits.
   When grown in the Algarve, the yellow cv. Flavicarpa is very resistant both to fungal-attack and to nematodes.

Peaches (alperches, pêssegos)

Of other crops grown on a plantation scale in the lowland regions, mention must be made of peaches, *Prunus persica* (Rosaceae), which flower in early April, the pink flowers produced on otherwise bare branches. Next to the apple, the peach is the world's most widely grown tree-fruit. It is probably a cultigen, selected in China and possibly derived from the wild *P. davidiana* there, and most probably was introduced to Europe in Roman times, being figured in the frescos at Pompeii. There are many cultivars grown for their fresh fruit and canning as well as juice, which can be fermented and distilled to produce a brandy. Dwarf forms are produced by budding shoots on to stocks of the Nanking cherry, *P. tomentosa*, also from China. In the Algarve, trees are grown mainly for fresh fruit, and to a lesser extent, for making jam, *doce de alperche*. In the annual cycle of fresh fruit grown in the province, peaches follow loquats (April), and apricots (May). Various cultivars provide fruit from the end of May until August.
   The nectarines ('var. *nectarina*') are cultivars with a smooth skin. A most successful very early nectarine grown in the Algarve is cv. Armking, distinguished by its yellow pulp and blood-red skin, and which can be harvested in May.
   In the Algarve, peach trees, which grow to 3.0 m and bear well until about twenty years old, are planted as two-year-old seedlings. These start to bear fruits the following year and come into full production within a further three years. They are pruned hard every December, and sprayed with 'Bordeaux Mixture' (principal ingredient copper sulphate) in early spring to reduce damage by peach-leaf curl (*lepra*, *Taphrina deformans*). Symptoms of this debilitating fungal disease include a thickening of the young leaves, which then discolour and pucker. In modern enterprises, watering of peaches is by a minispray system or, sometimes, a drip-feed at ground level, and is necessary throughout the dry periods from April onwards.

Cultivars

When grown in Portugal, peaches are often grafted on to seedlings, otherwise on to clones of almond such as 'INRA GF 677', depending on soil-type and whether nematodes (*Meloidogyne* and *Pratylenchus* spp.), and the bacterium *Agrobacterium tumefaciens* are present or not. There are three main groups:

(1) Peaches grown for fresh fruit (sub-divided into those with yellow or white flesh);

(2) Peaches grown for industrial (e. g. canning) purposes;

(3) Nectarines.

Cultivars are grouped according to the fruit maturation date. Within a group they vary in the colour, quantity, sugar content, size, perfume, and shape of the fruits, as well as the colour and texture of the pulp. A selection of those common in the Algarve, arranged by harvesting times is given below.

May

'Flórida', very early flowering, and vigorous; cv. Maycrest, matures one to two weeks later than 'Flórida'.

June:

'Amsden', the *Pessego de São João*; 'Morettini'.

July

'Redhaven'; 'Temporão de Alcobaça'.

August

'E. A. 158–45'; 'Maracotão'.

September

'Preto Carnudo'; 'Rosa'.

Pears (pêras)

Pears are derived from the Eurasian *Pyrus communis* (Rosaceae). Cultivars selected directly from this include Bartlett pears, very similar to those known to the Romans, and which are those most commonly canned. Many of those cultivars are self-incompatible and require cross-pollination. Pliny (AD 61–*c*. 113) knew 39 cultivars, sixteenth century Italy knew 232, and by 1831, 677 had been recorded in England alone, with over 1000 today. *P. communis* has been crossed with the Chinese *P. pyrifolia* (Chinese or Japanese pear, long naturalized in Japan), a species with apple-shaped fruits known for their good keeping qualities. The hybrids (*P.* × *lecontei*) include cv. Kieffer, which has enhanced disease-resistance, and may include the gigantic apple-shaped pears, cv. Passe crassane, sold in the Algarve markets in spring.

Cultivars

These are grafted on to rootstocks of *Pyrus communis* seedlings. Arranged by ripening times, the main cultivars planted are as follows.

June

'Lawson'

July

'Dom Joaquina', a small but very juicy pear common around Monchique; 'Delbard première'

August

'Rocha', very common in the Algarve and an excellent 'keeper'; 'Triomphe de Vienne'; 'Beurre Hardy'

September

'Duchesse d'Angoulême'

October

'Doyenne du Comice'.

November–December

'Beurre d'Aremberg'.

January–February

'Abate Fetel'; 'Passe crassane', a perfumed and juicy large apple-shaped pear grown mainly north of Lisbon in the Cadaval area.

Figure 185   Pear cv. Passe crassane.

Pineapple guava (feijoa)

The feijoa, *Feijoa (Acca) sellowiana* (Myrtaceae), named after the botanist Dom G. da Silva Feijoa, and now being grown on a small but increasing scale in the Algarve, is also known as the pineapple guava, but neither the fruit nor the tree bears much resemblance in appearance to either pineapples or guavas. *F. sellowiana*, the only known species of the genus is native in South America and is a small tree grown in many sub-tropical areas as an ornamental but more for its perfumed waxy dark-green fruits. They have a high sugar content, a somewhat tart flavour, and a taste reminiscent of strawberries. Cultivars are few, the best known being 'Mammouth' and 'Triumph'.

Plums (ameixas)

Plums (*Prunus domestica*, Rosaceae), are ancient cultigens grown on a small scale in the Algarve. They are very often grafted on to almond rootstocks, which avoids the necessity for watering. Cv. Mariana is one of the cultivars compatible with almond rootstock and the most successful grafts are made on almond trees 5–10 years old; fruiting starts the following year. Fruits ripen from the end of May until October.

Cultivars

There are many; examples are listed by ripening time.

July
cv.Rainha Cláudia.

August–September
'Coe's Violette', which has red pulp; 'Douradinha', a cultivar for drying (*ameixa para secar*).

Walnuts (nogueiras)

Known to the Greeks and Romans, the common walnut (*Juglans regia*) which probably originated in the area extending from the Carpathian Mountains eastwards to the Himalaya, has been grown on a small scale in the Algarve for centuries, mostly around Monchique, where there are still some trees. Today there are three main groups of walnut, as well as some hybrids:

(1) the common walnut, *Juglans regia*;
(2) 'black walnuts', which include *Juglans nigra* and *J. hindsii* (both originating in the eastern United States), were introduced to Europe in the seventeenth century;
(3) 'grey walnuts' and 'white walnuts', which include *Juglans cinerea* (from the north-east United States) and *J. sieboldiana* (originating in Japan), sometimes used in the Algarve as rootstocks.

The walnut, which is wind pollinated (*anemófila*), requires relatively cool weather throughout the year. Nevertheless, there is some plantation-scale culture around Silves, including selected cultivars from France; such trees come into bearing more quickly than those grafted on to *Juglans nigra*, *J. hindsii*, and older *J. regia* stocks.

Cultivars

'Arco', 'Rego', and 'Samil', all of Portuguese origin
'Amigo', 'Chico', 'Hartley', 'Serr', and 'Trinta', all from the United States
'Franquette' and 'Parisiense', both from France.

Walnuts can develop into trees up to 30 m tall with trunks up to 4 m in circumference. The durable timber of the tree is excellent for furniture, veneers, musical instruments, gunstocks, in ship construction, and for aircraft propellers.

In the Algarve most walnuts flower in March, a second group of cultivars later, and the nuts, which vary greatly in size and shape depending on the cultivar, are ready for harvest in August–September. For successful culture they need a comprehensive spraying programme to counter attacks of diseases such as walnut blight (*bacteriose*, *Xanthomonas campestris* (*X. juglandis*)), and insect attack by *Cydia pomonella*. Crown and root rots, caused by *Phytophthora cinnamomi* and *Armillariella* soil-borne fungi, are further causes of damage to the trees.

The oil from the nuts is used in paints but walnuts are better known as being excellent for eating (they contain up to 62 per cent oils). Their medicinal properties have been known since at least Pliny's time, who stated that 'chewing a nut fasting is a sovereign remedy against the bite of a mad dog.'

Soft fruits

Strawberries (morangos)

Of soft fruits, the only one grown on a significant scale in the Algarve is the strawberry, as rich in vitamin C as citrus. However, production is now declining owing to high labour costs combined with a relatively small-scale production. The large-fruited commercial strawberries — known in Spain as *fresones* — are hybrids (*Fragaria* x *ananassa*, Rosaceae), derived in the eighteenth century from a cross in France between *F. chiloensis*, growing wild along the American coast from Alaska to California and South America, and *F. virginiana* of the eastern seaboard. Other strawberries with smaller fruits — known in Spain as *fresas* — are derived from *F. vesca*, *F. muricata*, and *F. viridis*. What is referred to as the 'fruit' of all the strawberries is technically the swollen part of the base of the interior of the flower, the gritty 'pips' being tiny single-seeded fruitlets, the whole then rather like a fig turned inside-out (or vice-versa).

The major production area for strawberries in Portugal is Ribatejo, but in the Algarve, the strawberry is grown both at the *horta* scale for domestic use, and on *quintas*, which supply the local markets, hotels, and restaurants, while some strawberries are airfreighted to northern Europe. Strawberries in the Algarve are ready to harvest from the end of January onwards and a succession of cultivars is grown so that fruits are available until June. The crop is grown on level sheltered inland fields, an important region being the *concelho* of Silves. Production of strawberries in the Algarve uses a system similar to that practised in northern Europe, that is to say, rotating the areas where the crop is grown, so that new stock is propagated from runners every year and plants are replaced every three to four years. This ensures that stock remains virus-free and disease-resistant — mainly to *Phytophthora cactorum* and *Verticillium albo-atrum* — so maximizing the yield of quality fruits. An insect pest already serious in the Algarve is the thrips *Frankliniella occidentalis* (*tripe da California*), which attacks a wide range of crops including flowering strawberries.

Cultivars

Cultivars grown differ in some significant respects from those commonly seen in northern Europe in that the fruits themselves are remarkably firm and tasty. For use in jams, where weight is at a premium, very large 'mushy' varieties are grown towards the end of the season. The plants are grown in tunnels of black polythene and are watered by overhead spraying from March onwards. Cultivars which are tolerant of the limy soils that are typical of the most important agricultural areas of the Algarve have been selected. Annual yields are up to 8 t per hectare, though those of the Spanish strawberry plantations near Palos de la Frontera and Lepe, both in Huelva province, exceed 20 t per hectare.

The most common cultivars planted in the Algarve include cv. Chandler, cv. Douglas, cv. Sequoia, cv. Tioga, and cv. Tufts.

These differ from one another in ripening date and in the size, shape, and flavour of the fruit. All of them, especially cv. Tioga, a very vigorous and late-flowering cultivar with conical-shaped fruits, are common in Spain.

Raspberries (framboesas)

Shallow roots penetrating to only about 15 cm make raspberries, *Rubus idaeus* (Rosaceae), very difficult to grow successfully in the long hot summers of the Algarve. This also applies to the 'white-fruited' forms, which are the more lime-tolerant. Blackberries, loganberries, blackcurrants, gooseberries, redcurrants, and whitecurrants are equally unsuccessful in the Algarve.

Greenhouses (*estufas*)

Having a much greater effect on the landscape, particularly near Faro, are the polythene-covered greenhouses (*estufas*) used to raise early produce both for local

Figure 186 *Estufa* with melons.

Figure 187   Banana plant.

markets and for export. These were originally eucalyptus-framed, but are increasingly being replaced by galvanized iron.

Besides French beans, a number of other delicate crops are raised, for example bananas (*Musa* cvs.) and papayas (*Carica papaya*). However, these do not have any major economic significance, and the *estufas* are much more devoted to the production of early lettuce (*alface*) and, later on, early string beans, which are followed by melons and, most important of all, a succession of tomatoes which are available all year round without any heating. In the *concelho* of Silves, typical production in *estufas* is 60 per

Figure 188   Fruit stall at Portimão market.

cent tomatoes, 30 per cent melons, with green beans, peppers, and cucumbers making up the remaining 10 per cent. Tomatoes from the Algarve have a superb flavour and single fruits reach 1 kg in a well-grown specimen. In *estufas* they are grown up sticks in the northern European fashion, rather than by the outside 'ground culture' method practised all over the province.

Throughout the lowland zone of the Algarve, new crops are being tried and no list can be exhaustive. Watercress (*Nasturtium officinale*) is grown successfully at an enterprise near Quinta do Lago. In 1989 the entire crop, 800 t, was airfreighted to the United Kingdom. Smaller amounts have been grown for many years at other sites in the Algarve, and sold in the local markets, *mercados municipais*.

6

Serra and freshwater

Serra

The region known as the *Serra* covers elevated regions on different rock-types. To the west, the shales run out north of the *Barrocal* at Castelejo (p. 48) and, in the east, cover a wide area of hilly country to the Guadiana River at Alcoutim. On the whole the vegetation is monotonous and comprises mainly kilometre after kilometre of gum cistus (*Cistus ladanifer*) with brooms, *Genista triacanthos* (Plate 166)), *Lavandula viridis* (Plate 167), *Coronilla valentina* (Plate 168), *Ulex parviflorus*, and herbaceous species such as the small-flowered pink, *Petrorhagia nanteulii* (Caryophyllaceae). The straight roads south-west across the *Planalto* allow glimpses of the hills to the south, but little to detain the plantsman.

One hundred and sixty-nine plant species have been recorded from the Serra de Espinhaço de Cão, the hilly area west of Monchique, but the region is characterized by the restricted suite of *Cistus ladanifer*, *C. salviifolius*, *Genista triacanthos*, *Arbutus unedo* (the 'strawberry tree'), *Calluna vulgaris*, *Erica arborea*, and *Genista* (*Chamaespartium*) *tridentata*, which is gathered and sold as an infusion for a medicinal tea (*chá do calcaroleiro*). Plants tolerant of these conditions as well as those of the *Barrocal* include *Halimium calycinum*, *Lithodora prostrata*, here much more upright, *Lonicera implexa*, *Phlomis purpurea*, *Rosmarinus officinalis*, *Scilla monophyllos*, *Smilax aspera*, *Drimia maritima*, *Lavandula stoechas*, *Phillyrea angustifolia*, *Cytinus hypocistis*, *Pistacia lentiscus*, *Rhamnus lycioides*, and *Daphne gnidium*. Notable in spring are the spectacular broomrape *Orobanche gracilis*, *Orchis morio*, *Romulea bulbocodium*, and *Narcissus bulbocodium*, as well as the tiny Algarve endemic, *Linaria algarviana*.

Near the coast much of the flat area is often planted with rye or lupins but, over a great deal of this part of the Algarve, the *matos* have been replaced with plantations of eucalyptus, or are maintained for production of cork oak, the ground between the oaks periodically ploughed to encourage a fresh flush of herbaceous vegetation for grazing animals, or a poor crop of oats. This is then allowed to revert to *matos*, as in the south of Spain, so that a series of successional stages of the vegetation is maintained. This is carried out on such precipitous slopes that erosion is inevitable.

Likewise the steep areas around the *Barragem da Bravura* have been planted with eucalypts and, in spring after the heavy winter rains, the reservoir is a rich brown from the eroded soil swept into it. A national forest, *Mata Nacional de Barão de São João*, is planted with wattles (*Acacia* spp.), especially golden wattle, *A. pycnantha*, of

Figure 189 Plants of the *Serra*, I. 1. *Ulex parviflorus*; 2. *Geranium molle*; 3. *Coronilla valentina*; 4. *Anthyllis vulneraria*; 5. *Cytisus scoparius*; 6. *Euphorbia monchiquensis*; 7. *Lavandula viridis*; 8. *Teesdalia nudicaulis*; 9. *Bellis sylvestris*; 10. *Erica arborea*; 11. *Cerastium glomeratum*; 12. *Cistus crispus*.

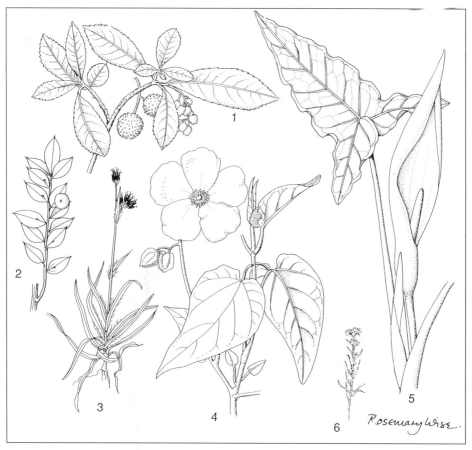

Figure 190   Plants of the *Serra*, II. 1. *Arbutus unedo*; 2. *Ruscus aculeatus*; 3. *Luzula campestris*; 4. *Cistus populifolius*; 5. *Arum italicum*; 6. *Scleranthus annuus*.

Victoria and South Australia, which has undivided (phyllodic) leaves and is an important source of tanbark. In addition, about one-third of the species recorded elsewhere on the Serra de Espinhaço de Cão are found there.

Because its height rises above the monotonous schist country, it is inevitable that the Serra de Monchique is the most visited montane area in the Algarve. The town of Monchique lies between the great syenite massifs of Foia to the west and Picota to the east. The geology of the surrounding area is complicated, but the land carries cork oak woodland as well as eucalypts and chestnut, and the first, when not too intensively managed, gives some idea of the original oak-dominated vegetation of the region.

The broad-leaved evergreen forest of the Mediterranean consists mostly of medium-sized trees, rarely over 18 m tall. *Quercus ilex* ssp. *ilex* (the holm oak) and its allies, including *Q. ilex* ssp. *ballota* (*Q. rotundifolia*) of the Algarve and the west Mediterranean, are circum-Mediterranean, but more abundant in the west, whilst,

Figure 191   Oak pasture on the *serras*.

Figure 192   Grazed pasture to left, regenerating *matos* right.

Figure 193   Reservoir and traditional tree-crops.

Figure 194   Barragem da Bravura murky with eroded soil.

Figure 195   *Mata Nacional de Barão de São João* with acacias and pines.

Figure 196   Cork oak woodland.

Q. suber, the cork oak (Plate 169), is found only there and then only on acid soils. *Quercus coccifera* of the *matos* is also exclusively west-Mediterranean, replaced in the east by *Q. calliprinos* (though this is considered conspecific by some authors). By comparison with the open *matos*, the forests seem species-poor, but it should be remembered that in an unmanaged state, there are continuous cycles of forest renewal with trees falling to produce clearings which become occupied by plants of more open habitats, before giving way to trees once more, so that the overall mosaic of vegetation would, indeed, be species-rich. Moreover, many of the woodland species are able to survive in these clearings and are also to be found in *matos* and even in the dwarfed vegetation of the coastal regions: these include *Lithodora prostrata*. Such examples may go some way to explain the origin of the species which are now to be found typically in man-made types of vegetation.

Most of the woodland evergreen species appear to retain their leaves for two years. *Quercus suber* generally loses part of its foliage before new leaves are formed, and there are differences in annual leaf production in the *Q. ilex* group, where a larger number of leaves is dropped in alternate years. Most plants flower in spring, though *Arbutus unedo* and *Smilax aspera,* like the carob, flower from late summer, the bell-like flowers of *A. unedo* usually being produced at the beginning of December, the fruits ripening at the end of the following summer, whilst the acorns of *Q. coccifera* take two years to mature. Not surprisingly, many of the evergreen species have been found to have greater drought-tolerance than their deciduous allies, for example *Pistacia lentiscus* compared with the deciduous *P. terebinthus*. Owing to the late-flowering of both *Arbutus* and carob, at a time when there are relatively few flowers of other species, many bees are attracted to them. Under experimental conditions, many of the evergreen forest species have been found to be more cold-tolerant than would be expected from their distributions. *Quercus suber, Q. coccifera,* and *Rhamnus alaternus* can withstand temperatures of −11°C, the *Q. ilex* group −15°C, *Myrtus communis* and *Nerium oleander* −8°C, and the carob −7°C.

Comparisons between the biomass of the high *Quercus* forest and *Q. coccifera garrigue* in southern France have shown that 150-year-old stands of *Q. ilex* there have 264 t per hectare, whereas a stand of *Q. coccifera* over 30 years old has only 37 t. Annual organic production per hectare was estimated at 1.7t for the high forest, accumulated as wood, and up to 7.8 t added to soil through litterfall. For the *garrigue*, the amount accumulated in wood was only 1.1 t per hectare per year, at least in a 17-year-old stand, and the annual litterfall was also much lower, at 2.6 t.

Compared with temperate deciduous forests, the Mediterranean evergreen forests have an inadequate fodder supply for many large mammals like deer, but wild pigs are reasonably common, and hunting for them, rabbits (*coelhos*), as well as for partridge (*perdiz*) in season, is carried on in the more remote and mountainous regions of the Algarve. Owing to disturbance of their habitat, rabbits are now found very infrequently in much of the *Litoral*, and they do not proliferate in the *Barrocal*. Birds are indiscriminately shot all over the province, even in the coastal protected areas. Quail (*codorniz*) and small birds similar to them, are killed in large numbers. Since 'hunting rights' are an intrinsic part of the Algarvian way of life, the law permits hunting in most 'country' areas as well as inside the designated hunting zones (ZCT,

Zonas de Caça Turística, of which 295 were defined in Portugal in 1992), with the proviso that it does not take place too near occupied dwellings.

In the Mediterranean as a whole, the principal forest product is wood for fuel. The major products in the Algarve are cork (*Quercus suber*), timber of various species (mainly eucalypts and pine), and the fruits of *Arbutus unedo* for distillation into *medronho*. The applications of derivatives of these trees are described later in this chapter.

Near Monchique, the understorey of cork oak woodland includes *Calluna vulgaris, Erica arborea, E. australis, Arbutus unedo, Cytisus scoparius* (Plate 170), *Euphorbia monchiquensis* (Plate 171), *Genista tridentata* (Plate 172), *Lithodora prostrata, Phillyrea angustifolia, Daphne gnidium* and the laurustinus *Viburnum tinus* (Caprifoliaceae, Plate 173), *Lavandula stoechas, Scilla monophyllos, Cistus salviifolius, C. populifolius* (Plate 174), *C. crispus* (Plate 175), *Arum italicum, Myrtus communis, Paeonia broteroi* (rare, Plate 176), *Ruscus aculeatus* and the liane *Smilax aspera*, with *Umbilicus rupestris* on exposed rocks. By the end of June, at altitudes up to 600 m around Monchique, foxgloves, *Digitalis purpurea*, send up massive spikes to 1.5 m, variable in flower colour, thriving equally well on roadside slopes and in nearby *mato*.

Cork

Cork oak, *Quercus suber*, is widely grown in Portugal, mainly in the province of Alentejo, which is home to one third of the world's cork trees: 750 000 ha of the world total cork plantations of 2 270 000 ha and more than a half of the world's annual cork production of about 400 000 t, in recent years ranging between 207 000 t in 1987 and 255 000 t in 1984. Portugal, the most wooded of all Mediterranean countries (39.8 per cent cover), has about 3 million ha of forests, so that cork represents about a quarter of them. Often the plantings are irregular, and such is the value of the product that even single trees are grown commercially. In the Algarve, the most important centres of production of cork oak are in the *concelhos* of Loulé, Monchique, Silves, São Brás de Alportel, and Aljezur in that order, and the total afforested area of cork oak in the province is 42 000 ha.

Cork oak is planted in a belt (*c.* 30°–50°N) which extends into western Spain spreading southwards to Morocco and Algeria, and eastwards in a narrow coastal strip to Italy and Sardinia, and into the adjacent hills up to about 400 m altitude. A secondary belt of cork oak is planted between 30°S and 40°S in parts of South America, South Africa, and southern Australia. It is grown where annual precipitation is in the range 400–1000 mm and where the temperature does not go below –5°C. In Portugal, cork oak replaces *Q. ilex* ssp. *ballota* on poor, acid (pH 5–6), stony soils, where its slow growth yields the finest quality cork.

The cork of commerce, *casca de sobro*, is derived from the outer bark of cork trees. Certainly important in the fourteenth century, as the records of Évora show, and until the early 1900s, the inner layer of the cork oak was worth more than the outer bark for the tanning industry. In 1815, Lisbon exported to England 13 700 *arrobas* of cork (as tanbark), and nearly three times that amount in 1818. But, as the cambium was removed with the inner layers, the trees died, and the wood was then used for charcoal production. In turn, this requirement for tanbark from cork oak arose in part from the

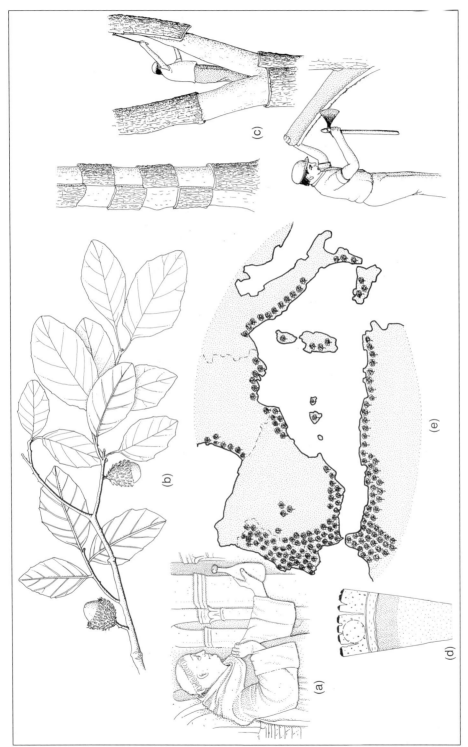

Figure 197 Cork (*cortiça*) and its uses : (a) early stopper, (b) fruiting branch, (c) cork-stripping, (d) cross-section of trunk showing bark and position from which cork is cut, (e) distribution in the Mediterranean region. (Redrawn from Amorim 1989.)

Figure 198　　Stripped cork oak.

Figure 199　　Recently stripped and numbered cork oak.

severe depletion in Spain as well as Portugal of the once extensive stands of sumach (*sumagre, Rhus coriaria*) and of myrtle (*Myrtus communis*). In the Algarve in the nineteenth century, sumach was important for the tanning of hides, *curtimentas*, and cork oak was little used there then. The rise of the synthetic chemical industry at the turn of the century led to a great decline in tanbark production from cork oak and the

Figure 200   Cork stack.

Figure 201   Cork transport.

use of cork as tanbark was then made illegal in Portugal. Also cork oak has always been of importance for its acorns, the access rights to which are known in Portugal as *montanheira*, for use as pig feed; by 1961, Portugal's cork acorns were being converted annually into 18 000 t of pork.

Uses

Cork does not have a homogeneous structure. The cells are 14-sided polyhedra separated by gas-filled intercellular spaces, forming layers. These are divided by alternating sheets of waxes (5 per cent) and suberins (45 per cent), and the entire structure is made porous by radial channels. Other chemicals contained in the vegetable structure include a wide range of tannins (6 per cent), polysaccharides (12 per cent), and lignins (27 per cent). The use of cork for making stoppers for wine vessels (*amphorae*) was mentioned by Pliny, Horace, and Cato, but in those times the stoppers were not dense enough and had to be sealed with resin. The modern industry may be said to have taken off when high-quality cork stoppers were used for champagne bottles in the Benedictine Abbey at Hautvillers (Marne, France), where the first sparkling wines, and glass bottles in which to keep them, are recorded in the seventeenth century. Whilst the widespread use of small-necked glass bottles speeded the introduction of cork stoppers, Portwine and Madeira had long been exported in casks stopped with cork.

In 1892, a method was developed to convert ground cork waste and low-grade cork wood into a range of composition corks and corkboard, but real progress in the application of the process was not made until after the First World War. With the new technology, some 50 per cent of the material, which previously had been wasted, was readily convertible into a valuable commercial product by granulating it, mixing with a binder and then compressing and baking it for 20 minutes at 380°C. The method is still used today. Between 1925 and 1960, annual production increased three-fold. In more recent years, further high-value applications of the outer cork layers have been developed, including slitting machinery to produce high-quality cigarette filters. Grades of cork are made to provide sound insulation in hi-fi studios, with grades of lower quality being produced for domestic use for inclusion between double-brick walls. Various compounding methods to make decorative floor-, wall-, and ceiling-tiles, coated and bonded tiles with rubber backing, mainly for use in flooring, have been devised. Other grades suitable for insulating refrigerated ships and trucks are manufactured and there are special grades for bearings used in vibrating machinery, and others for highly efficient filters for a wide range of industrial processes. Other uses of cork include floats, shipboard and quayside absorbent bumpers, engine sealing gaskets of many types, and expansion joints widely used in buildings, bridges etc. Cork is also the key component of shuttlecocks which are mainly exported to Pakistan, India, and Indonesia.

Harvesting and production

The cork oak trees are grown for about 25 years (eight to ten in other countries), after which the first layers of cork are stripped, starting from the lower trunk. It is removed with a special axe, similar to that of the medieval executioner. The material is removed from the trunk of the tree (and then from the upper main branches) by making vertical slits up to 2 m long. After some horizontal incisions have been made at either end, taking care not to damage the cambium layer, the same axe is inserted behind the cork and the curved strip levered away from the trunk. Skilled workers can cut 150–400 kg of cork a day. The first cutting, of 'virgin cork', is usually of low quality, but further removals of the cork are made at about nine-year intervals in the Algarve, until the tree is finally exhausted at about 170 years, when it is grubbed out, sold for firewood or charcoal and replaced.

The cork layers are from 2–5 cm in thickness. After the cork sheets are removed from the trees, the curved sections are sorted into various qualities and stacked nearby for some months, when they become almost flat. Then, the lower qualities are transported to factories which deal with agglomerates, cork board etc., whilst the better material is taken to other plants where it is slit, coated, bonded etc. After the trees have been cut, they are code-marked with a date on the trunks which are then a vivid red colour, but dull rapidly as the cork, i.e. bark, regenerates. Cork is very variable in its growth, and even trees under apparently similar conditions in the same plantation vary over a wide range. In the Algarve, cork plantations are now generally grown with all cover removed from the understorey, and are intensively managed. Excellent examples of cork forests can be seen around Nave. A cork cartel, administered by the Portuguese authorities, ensures that material reaches the markets at the highest price in quantities which approximate to demand.

Medronho

To the north-west, north, and east of the Monchique massif are relict trees of *Quercus canariensis* (Plates 177 and 178), an evergreen species of the western Mediterranean (not the Canary Islands) in its only site in Portugal. On the slopes of Picota is found *Myrica faya* (Myricaceae), probably not native as it is in the Azores, whence it has been introduced to Hawaii, perhaps by the Portuguese as a fruit-crop in winemaking, and where it is an invasive pest. After the destruction of the oak forest, the first phase is dominated by *Arbutus unedo*, accompanied by *Cistus populifolius*: destruction of this leads to the *Cistus ladanifer* assemblages typical of the shales. However, *A. unedo* is found commonly only above 400 m.

Figure 202 *Medronho* plain (left) and with honey (right).

The best-known *aguardente de medronhos* of the Algarve is that made from *Arbutus unedo* (Plate 179), but another kind is produced at Pera from strawberries. The distillation of dried figs yields the potent *aguardente de figos*, pure alcohol, the red colour of the figs being imparted to the liquor. To produce the *medronho*, the ripe fruits are harvested between September and November. A number of processes, mostly of a primitive kind, are used, and the product is very variable. However, most involve packing the fruits into open-topped wooden barrels (*vasilhas*). Water is added and fermentation begins; to prevent a second fermentation — to vinegar, a layer of mud is put over the top to exclude oxygen. In January, the mass is transferred to a copper still and boiled with a serpentine tube fitted over the top of the kettle. As the tube lies in cold water, the evaporated alcohol condenses and is collected. Better qualities of *medronho* undergo multiple distillation. The final product can be 90 per cent proof and 15 kg of fruits yield up to 2 l of spirit. This is very much higher than that in the production, for example in Austria, of *Vogelbeerschnaps*, distilled from the fruits of the rowanberry *Sorbus aucuparia*, but is similar to that of *slivowitz*, produced in Yugoslavia from ripe plums. In the Algarve, most *medronho* which is not drunk in the household is sold to licensed bottling factories, and there it is mixed with pure alcohol (ethanol) and various essences. Export is forbidden, allegedly because the drink contains too much methanol (up to 100 g per litre).

Another, although relatively minor, use of *Arbutus unedo* is in the production of jam. Whilst the somewhat bland taste of the fruit makes it generally unacceptable for eating fresh, it, rather surprisingly, produces excellent jellies and jams, and its high pectin content gives rapid setting without additives. The product has a very pronounced 'strawberry' taste.

Pine

Maritime pines have been established widely in the hills as well as in coastal zones for timber and turpentine. It is estimated that there are 1 248 000 ha of *Pinus pinaster* (*pinho bravo*) in Portugal. For comparison, there are only 30 000 ha of chestnut (see below). *P. pinaster* is the richest Mediterranean pine as far as resin production is concerned, Portugal's annual production being 108 000 t, though resin from *Pistacia terebinthus*, and a gummy mastic from *P. lentiscus*, were prized in classical Greek times. In the Algarve, records from the early nineteenth century show that resins (*resinas, almécega, mástique*) were important articles of commerce. Here they were obtained not only from *P. lentiscus*, but also from *Phillyrea angustifolia*, and, less obviously, from almond, apricot, and plum trees.

In antiquity, pitch from pines was an important trading commodity and south-west Spain exported it. Pitch was, and is still, used mainly for the caulking of ships and for other buildings purposes including house construction, whilst resin has a wide variety of economic uses including the production of wines, better known in the *retsinas* of Greece. The resins are tapped by making shallow diagonal grooves, and collecting the exudate. After four or five years, the trees are rested. Pitch from trees is being replaced by a wide range of products from the distillation of crude oil.

South-west Europe was also considered to have good supplies of ship-timber. Fir (*Abies* sp.) was considered best for ships in the classical period, whilst today pines are

Figure 203   Resin tapping, *Pinus pinaster.*

Figure 204   Fishing-
boat repairs with
pine, Portimão.

used to repair many of the fishing boats at Portimão and other Algarvian ports, the
timbers coming from the Serra de Monchique. The parts of the boats which need to be
especially strong, such as the keel, are made from *Pinus pinea* (*pinho manso*), with the
rest of the hull being *Pinus pinaster*. In both cases, pine is favoured because it is
available in long straight lengths free of knots, and, when used in boats, lasts for up to
40 years. The large fleets of caravels, *naus*, and galleons used by Henry the Navigator

Figure 205 Fishing-boat repairs with pine, Portimão.

Figure 206 Traditional pine cart.

(*0 Infante*) had resinous pinewood in the keel and stem, and other parts where water penetrates, whilst the ribs (*cavername*) of the ships were usually made of cork oak and holm oak, both of which are water-resistant. Other parts of the ships, where water did not reach, were made from *pinho manso*.

Pine is also used for the famous Algarvian mule carts, including the wheels, though these were formerly made from oak, *azinho*, for the hubs and rims, with Brazilian hardwood for the spokes. The wheelwrights who make and repair these carts are specialists, and are known as *carpinteiros de carroças*, but few remain and are found now only in Alcantarilha, Ferreiras, Portimão, and Santa Barbara de Nexe. The *carroças* were part of the traditional integrated agricultural system of the Algarve, but there is little place for them in an era of increasing motor traffic, although many of the older farmers, as well as gypsies, still favour them. The mules themselves are kept in buildings adjoining the farmhouse and the manure, *estrume*, is used on the crops. Today, these carts are still used to convey the grapes to the cooperatives after the harvest in autumn, or to visit the rural markets which are held

Figure 207   Cartwheel.

in towns and villages monthly, when many itinerant traders are to be seen selling their wares.

Eucalyptus

There are over 600 species of *Eucalyptus* (gums, Myrtaceae) native to Australia and the eastern part of the Malay Archipelago. They are evergreen, usually with distinctive juvenile phase differences in the shape, colour, and positions of the leaves. When in bud, the flowers are covered with small lids (opercula) that fall after the flowers open. The petals and sepals are vestigial and the colour of the flowers comes from the filaments of the stamens. These filaments are usually pale-yellow or white, but can be bright red, as in the Western Australian *E. ficifolia*, cultivated as an ornamental in the Mediterranean. The seeds are tiny and are produced in large numbers in each woody fruit, which may remain on the tree for many years. Over 200 species have been planted in different parts of the world, largely for timber. Some eucalypts have transformed the landscape of certain countries, for example parts of the highlands of East Africa, Sri Lanka, California, and much of the Mediterranean region (1 million ha out of the world's 3.7 million ha of eucalyptus plantations) including Portugal. Because gum leaves are held vertically rather than horizontally, an adaptation to the intense radiation in Australia, light is reflected in a way different from that in native trees, giving the landscape an entirely 'new' look.

In Portugal, only a few species have been planted for timber production. In the Algarve, there is the frost-resistant *E. globulus* ssp. *maidenii* of south-east Australia

Figure 208   *Eucalyptus globulus* plantation near Marmelete.

Figure 209   Terraced *Eucalyptus globulus* with citrus in valley.

Figure 210   *Eucalyptus globulus* coppicing.

(stalkless fruits) as well as *E. camaldulensis*, (stalked fruits, Plate 180) which have flowers in groups, and, much more importantly, *E. globulus* ssp. *globulus* (*eucalipto vulgar*, Plate 181), which is that most commonly seen in plantations, and has the flowers borne singly, a condition rare in the genus. *E. globulus*, the blue gum of

Victoria and Tasmania, introduced to Italy in 1803 and to Portugal in 1829, is by far the most widely cultivated species in the world, and is naturalized in many places such as California. The first eucalypt plantations were established in Portugal in 1875, but extensive planting, induced by changes in papermaking technology and lax planning controls, began only in the 1960s. By the 1980s, the realization of the economic advantages (rapid growth rate) led to further major increases in planting of eucalypts. Recent studies in Portugal have shown that *E. globulus* growing under ideal ecological conditions can attain a height of 28 m (averaging 22 m) within ten years of planting, compared with an average height attained under ideal conditions by *pinheiro bravo* of 22 m after 50 years. No wonder then that, by 1990, afforested areas of eucalyptus (more than 95 per cent being *E. globulus*) already exceeded 500 000 ha, more than 15 per cent of the total forest lands of Portugal.

Often, as in Ethiopia for example, eucalypts are grown as a crop for firewood, with the flowers providing good bee forage, yielding a magnificent aromatic honey. They are an important source of eucalyptus oil, especially in Spain. In the Algarve, eucalyptus leaves are gathered for plunging into bath-water, which releases the essential oils. The juvenile foliage is used in floristry, but it is the timber with its short fibres which provide the most desired properties for the production of paper pulp. Since the timber is pale-coloured, it requires little bleaching to produce 'white' papers of high-value by the acid-bisulphite method, which removes lignins from the cellulose. 'Kraft-quality' papers are produced by a different process which uses an alkaline solution of sodium sulphide and caustic soda to break down the wood-pulp and, because recycling is possible, waste residues can be all but eliminated in a properly designed and operated plant using this process.

Other uses of eucalyptus timber in the Algarve include the production of beams for roof trusses in houses, scaffold poles, and generally for supporting structures in the building industry, structural beams for *estufas*, for railway sleepers and other strong flexible blocks to support heavy vibrating machinery, and telegraph poles, but over 90 per cent of the total production (1990, 1.5 million t *pasta de papel*) is now exported as pulp, although here increasingly with higher added value as paper.

Production

On the best soils in Portugal, a sustained annual production of wood of 50 m^3 per hectare of eucalyptus is achieved, about five times that from *Pinus pinaster*. Eucalyptus pulp production in 1989 by only one of the major Portuguese producers, Soporcel Lda., a company which is both grower and pulp producer, was 387 000 t with a sales value of 43×10^9 escudos (approximately US\$0.3 $\times 10^9$). Since Soporcel's afforested area increased from 14 500 ha in 1987 to 32 300 ha in 1989, it is evident that the company's production of eucalyptus will be of the order of 1 million t per annum within five years, even without further plantings. Another major producer, Celbi, with an annual production of 255 000 t of pulp, has already obtained Government consent to extend this to 600 000 t. The same company is far advanced in the production of genetically superior *Eucalyptus*, promising far greater yields. In Portugal there are eight factories owned by four companies producing wood pulp, mainly eucalyptus. That of Portucel at Viana do Castelo is an integrated pulp and paper producer. In 1985, the pulp factories together processed about 7 million m³ of wood, of which 20.2 per cent was pine, and 79.8 per cent eucalypt. Portugal's most important export markets by far lie in Continental Europe. *Eucalyptus* (with *Acacia* spp. — mainly *A. dealbata, A. pycnantha, A. longifolia*, and *A. melanoxylon*), is grown in a wide belt along the coast of Portugal from Porto in the north, down to the Algarve. The location of the plantations is as much a question of nearby large rivers to provide water for processing and for waste disposal as it is of the soil and climatic conditions necessary for the crop. Not surprisingly, therefore, five of the eight of Portugal's major eucalyptus factories are near, and are discharging their wastes into, the estuaries of rivers Vouga, Sado, Caima, and Tejo (two), and the remaining three factories are discharging their wastes directly into the ocean. Liquid effluent, all contaminated to some degree, is produced at a rate of 55–170 m³ per tonne of treated material. Residual solid wastes, mostly rather more toxic and dangerous, already exceed 71 000 t annually from the pulp and paper industry of Portugal, though oxygenation ('pre-whitening'), a process developed by Celbi, may reduce the hazard.

In all these places, the mean annual temperature is about 15°C, and mean annual rainfall between 800 mm and 1000 mm. Whilst in much of Portugal marginally productive, and relatively small, vineyards, olive groves, and pasture have been con-verted into large eucalyptus plantations, most of the eucalyptus in the Algarve has been planted on the hilly slopes in the *Serra* and *Barrocal*, for example around Monchique, land which would support few alternative trees other than oak and pine.

In the Algarve, the production method is to plant seedlings in a dense matrix, and to coppice them after an average of eight years, by which time *E. globulus* stems attain a diameter of at least 15 cm. Regrowth occurs quickly and reliably from the stumps and, during the following year, the new shoots are thinned to the three or four strongest, after which these are again coppiced at about eight-yearly intervals. Under less favourable growing conditions, the coppicing cycle may be extended up to twelve years, whilst the cycle can be as short as five years in ideal ones. After about three or four successive cuts, some 30 years in all, the plantations are replanted. As fires promote seed germination in eucalyptus plantations, they have been used in the management cycle, where, after removal of old trees, a few mature trees are left on a scattered grid through the plantation, and then deliberately burnt. As a result seeds are spread over a large area from a great height, in turn rapidly providing more seedlings and a new cycle, with a minimum of labour input. Trimming of the felled timber is carried out in the plantations, and all the side branches, bark and other litter is left where it falls. Finally, the poles are cut into 2 m lengths for transport by road to the processing plants.

Environmental effects

Environmental 'green' groups are vocal in Portugal in objecting to eucalyptus. Doubts raised centre on the following problems.

Increased risk of fires

These arise from the large volume of inflammable litter, including dried leaves, and from the oil content of the tree. Recently, Portugal has recorded about 2000 major forest fires annually, though in 1991, by mid-August, over 18000 forest fires were registered, with the devastation of 120 000 ha. In the Algarve, in the same period, there were 649 fires damaging 6364 ha of forest — as well as a considerably larger area of *mato*. It remains unclear whether or not these fires are exacerbated by the increased planting and management of eucalypts. They probably are, although the total financial and social consequences cannot be assessed accurately yet.

Effect on water availability

The objection is that the rapid growth rate of eucalypts leads to a lowering of the water table. This is probably so, although scientific evidence is scanty. None the less, so far this water depletion does not appear to have had any serious consequences overall, since eucalypts are generally grown in areas of high precipitation. In the Algarve, they are mainly in hilly areas, where rainfall is high and there are no other major competing demands for the available supply. The competition for water from eucalypts is recognized to some extent, since it has long been illegal (*Decreto-Lei N° 28 039*, 14 September 1937) to plant these trees within 20 m of other cultivated land.

Effect on soil nutrients

It is argued that the extraordinarily rapid growth rate of eucalypts will irreversibly diminish the minerals usually found in the soil. Superficially, this appears to be so in the sense that vegetation found under eucalypt canopies is not species-rich. However, there may be other substantive reasons for this, for example, the effects of litter arising from coppicing.

Soil erosion

This is already a serious problem, but may be reduced by ploughing along lines of equal altitude, never down the slope, and by careful maintenance.

Monoculture

Potentially this is the most serious difficulty of all, as it is entirely possible that, at some stage, severe, and uncontrollable, pest problems will follow in the wake of these plantings. Already there is evidence that eucalypts are being damaged by serious fungal diseases (*Armillariella* sp., and *Rosellinia necatrix*), and by fatal attacks of the coleopteran *Phoracantha semipunctata*, *broco do eucalipto*, which is already affecting up to 12 per cent of the plantings.

Figure 211
Maturing roadside
Eucalyptus globulus.

Commentary

Without question, the plantings of eucalypts on a large scale can have only negative effects on flora, fauna, fire hazard, overall nutrient soil balance etc. But to be set against these, there are the undoubted economic advantages in the short-term provided by profitable exploitation, with over 80 per cent of the production going to export markets. On balance, it is imperative to ensure that a scientific programme of plantings and subsequent management is designed and implemented; such might include other species, e.g. pine, oak, and chestnut. In areas where cork oak can be grown well, that species should be favoured, since even on economic grounds *Quercus suber* provides a superior return, and with little of the adverse environmental impact of eucalypts. In the Ribatejo area, it has been calculated that cork oak yields about US$700 per hectare per annum, against a yield from eucalyptus of only US$440 per hectare per annum. Such calculations do not take into account the 'clean-up costs' associated with the argument that 'the polluter should pay'. With such considerations, eucalypts would be a much less attractive proposition than they are now.

Monchique

Whilst the earliest agricultural settlement in Portugal has been dated to 5000–3000 BC in the Lisbon area, the beginnings of agriculture in the *Serra* were at the junction of the syenitic and shale zones some four thousand years ago, at Caldas de Monchique. By the eighteenth century there was a thriving economy, importing grain and exporting chestnut wood, dried fruits, and vegetables. Many of the vineyards of that period later fell to the ravages of blight and, by 1915, of *Phylloxera* (*filoxera*), and olives and some figs were substituted, though dessert grapes are still grown. Caldas de Monchique today gives the impression of having been a genteel spa town, exactly what it was at the turn of the century. It is a site with extensive, but poorly excavated and restored, Roman remains, and from its springs comes a high quality mineral water in economic quantities, just as in Roman times. A large bottling plant and a thermal bath treatment hospital reflect the town's continued significance today.

In 1824 the King, Dom João VI, granted rights to the land on the upper slope of Foia to the people of the town of Monchique. The forest cover which had persisted until then was rapidly removed, so that the modern tree-line at about 600 m is much lower than originally. Today, there is increasing development for housing and the only forest at the summit is one of forbidding transmitting aerials. On the flanks of the mountain, there are areas of *Cistus* and *Erica* species, with stands of *Rhododendron ponticum* ssp. *baeticum* (also recorded from the slopes of Picota, and now becoming increasingly rare, Plates 182 and 183), in the wetter sites. This rhododendron is restricted to acidic soils in southern Spain, the Serra do Caramulo (Guarda, near Oporto) and Monchique. *R. ponticum* ssp. *ponticum* is native in south-east Europe and south-west Asia, but it is widely naturalized throughout northern Europe, sometimes becoming a serious pest on acid soils as in forestry plantations in southern England, where it was introduced in 1763. In the last interglacial period it was native over much of Europe, for example in Ireland, and ssp. *baeticum* is thus considered a relic of that wide distribution. Its evergreen foliage is poisonous. On Foia, it flowers in June and is

Figure 212   Pico de Picota from Monchique town.

found amongst bushland of *Cistus salviifolius* and *C. ladanifer*, *Erica arborea*, and *E. australis*, bracken, *Ruscus aculeatus*, *Arbutus unedo*, *Daphne gnidium*, *Drimia maritima*, *Lavandula stoechas*, and *Tamus communis*, along with the prickly *Ulex minor*.

In the grazed grasslands near the peak, there are rocky corners with *Romulea bulbocodium* (Plate 184), *Saxifraga granulata*, *Scilla monophyllos*, which is very variable in size and colour here, *Luzula* sp., *Arisarum simorrhinum*, *Paeonia broteroi*, *Sedum nicaeense*, and *S. brevifolium*, besides the cistuses and heathers. The area is of more interest to bryologists, and minute angiosperms include *Teesdalia nudicaulis* (Cruciferae) and *Scleranthus annuus* (Caryophyllaceae), both these plants being more typical of regions further north in Portugal and the rest of Europe, whilst rather pathetic hawthorns, *Crataegus monogyna* ssp. *brevispina,* are the only tall plants in sight.

More rewarding are the slopes of Picota, although in recent years the forest paths have become more readily accessible to four-wheel drive vehicles, and tourist pressures have seriously damaged the vegetation, as has been the case on Foia since the 1970s. The track from the Alferce road leads up through chestnut (*Castanea sativa*) coppices, the old stools of which grow to 1.5 m width. The floor under the trees is dominated by *Asphodelus ramosus*, occupying much the same position as bluebells do in a coppice wood in the British Isles. *Euphorbia characias* (Plate 185) is the counterpart of *E. amygdaloides*, and both primroses (*Primula vulgaris*) and foxgloves (*Digitalis purpurea*) familiar in northern Europe are to be found here.

Chestnuts (Plate 186), *castanheiras*, which are not native in the Algarve, have been grown since the 1820s over a wide area on the slopes around Monchique, but much there was lost owing to blight, *Endothria parasitica*, at the beginning of this century. Scattered among these trees are small numbers of walnuts, which are planted for their mature nuts, rather than, as in England, for picking under-ripe and preserving. The chestnuts rival the better-known Spanish nuts in both size and quality, but the trees are cultivated much more for their wood. For this the trees are coppiced, so that resprouting occurs in a process which is repeated several times over about 50 years. The wood is prized for making beds and other furniture, where its grain and colour as well as

Figure 213
Chestnut coppice on
Picota.

Figure 214   Inside chestnut coppice woodland.

strength, are valued. It is widely used for making ladders, but eucalyptus and pine have replaced much of the chestnut in making furniture. Chairs made from chestnut sometimes have backs of twisted reedmace fibre, *Typha latifolia* (*tábua larga*) and *Typha angustifolia* (*tábua estreita*). The 'green' wood is much used for making wine-barrels. Chestnut poles have a range of uses on the *hortas* and *quintas*, including fencing, tool handles, and staking. From the chestnut poles cut on Picota in late winter, thin strips are split and then made into baskets for carrying fruit and vegetables.

Figure 215
Chestnut stool and
coppice shoots.

Figure 216 View
from Picota towards
the sea.

ʹAbove pine, chestnut, and eucalypt plantations on Picota is an open shrubland of the rather mat-forming, heavily galled *Quercus lusitanica* (Plate 187) on the rocks with *Q. coccifera* on the drier sites, with *Erica australis* (Plate 188), *Halimium calycinum* (Plate 189), *E. arborea, Bellis sylvestris, Romulea bulbocodium*, its flowers with a wide variation of colour and size, and *Calluna vulgaris*. Other *mato* species like *Juniperus phoenicea, Cistus salviifolius, Scilla monophyllos*, and *Lithodora prostrata* are there, and, also found on rocks, *Centranthus calcitrapae* (Valerianaceae, Plate 190), *Sedum album, S. brevifolium*, and *S. forsterianum* (Plate 191). The co-existing *Sedum* species have somewhat different microhabitats, *S. brevifolium* being largely found on the southern slopes, *S. forsterianum* only in sheltered areas of vegetation, and *S. album* largely on the exposed shallow soils. The vegetation is at its most floriferous early in May, when dwarf forms of *Cistus ladanifer*, both 'blotched' and 'unblotched', *Cistus crispus*, and the handsome *Tuberaria lignosa* (Plate 192) with its butter-yellow flowers, are at their best. On shallow soils and in crevices are *Allium massaessylum*, *Arenaria conimbricensis*, *Cleome viscosa* (Capparidaceae), and *Dipcadi serotinum*, while the woods are greening with bracken and at path edges are plants of *Simethis planifolia* (Liliaceae), its white flowers remarkable for their hairy stamens.

In the region are two other interesting plants, a large buttercup, *Ranunculus bupleuroides*, otherwise unknown in the Algarve, and the carnivorous *Drosophyllum lusitanicum* (Plate 193) growing, with *Limodorum abortivum* (Orchidaceae, Plate 194), in cork oak woodland and along nearby roadsides. First recorded in Portugal in 1689 by Tournefort, this unusual plant is known from western Portugal, southern Spain, and the northern tip of Morocco, with a few inland sites away from the sea-mists held to be important in controlling its distribution. It often occurs close to zones of earlier volcanic eruption. Unlike most carnivorous plants, it is found in very well-drained soils which are baked dry through the summer. It is salt-tolerant, and near Monchique can be found in disturbed ground, with the building boom in Spain and Portugal providing new habitats for it, although it is sometimes found in stable habitats under *Pinus pinea*, as near Algoz. As the vegetation grows up, *D. lusitanicum* scrambles up with it, the base of the stem retaining the old dead leaves, giving the whole system the appearance of a contorted flue-brush. Stem sections of 1.1 cm are known to have been six-to-seven-year-old plants, and some of the Monchique specimens are stouter, although sometimes the plant behaves as a biennial.

For a carnivorous plant, *D. lusitanicum* has a rather extensive root-system, though this is not so developed as would be expected from its habitat and the fact that it functions as an insectivore throughout the dry season. Moreover, the glistening stalked tentacles have only a vestigial cuticle to prevent overall water loss. It is difficult to explain the water-balance of this fascinating plant. Insects are attracted to *D. lusitanicum* by a strong honey scent, and become entangled in mucilage on the leaves, which are coiled facing outwards, whereas in the related *Drosera*, they are coiled facing inwards. The mucilage is produced by stalked glands which also produce some proteolytic enzymes. Most of the digestion of the insect seems to be by the stalkless glands on to which the animals fall between the stalked glands, which do not move to entrap the prey, unlike those in *Drosera*.

Other products of the Serra

Typical of the *Serra*, including around Monchique, is the production of charcoal (*carvão*), in former years generally associated with the production of tanbark from cork oak which led to early death of the trees. A stack of wood in the form of a flattened hemisphere, about 4 m tall and 8 m across, is allowed to burn slowly for a week or so. To maintain the rate of combustion, holes for air are drilled into the pile. The charcoal is important as it is used for the large number of open fish grills. Best known perhaps, to the local people as well as tourists, are the grilled sardines (*sardinhas assadas*), always served with fresh-baked bread on which the sardines are placed to absorb their oil, and 'com salada e batatas cozidas' (a salad of lettuce, sliced tomatoes, and large onions, with boiled potatoes) and washed down 'com vinho tinto' (with local red wine).

Honey (*mel*) has already been mentioned in connection with eucalyptus, but the collecting of honey is a very ancient practice, because it was used for sweetening before sugar became available generally. Even today, it is produced all over the Algarve, though this is very much a cottage industry, and thus meaningful production figures are not available, since the excellent products are consumed directly by those making them, or disposed of to local markets. In 1990 the national production was said to be some 3500 t. but the industry in the Algarve appears to be in decline, largely with the eclipse of the smaller farms, owing mainly to pressures from the tourist industry, so that much of the honey sold today in the Algarve is imported from Australia.

Another important product of Monchique derives from its very bedrock, syenite, which is dynamited from quarries in the hills, the largest being at Nave. This is the most important mineral working in the Algarve and output has recently increased to meet Japanese demand. The blue 'granite' is blasted off the mountain-side in large blocks,

Figure 217 Syenite quarry at Nave (old Navête).

Figure 218 Chiselling syenite, Nave (old Navête).

which are then broken down by an extraordinarily accurate and efficient technique with a hammer and chisel finally into blocks (11 cm cubes) used for paving. This work, called '*calçada Portuguesa*', or '*método da calçada*', has been known since at least Roman times, when the blocks were used for roads. Many of the Algarve's older roads are today covered with asphalt, but the granite blocks are still widely used in paving and many larger ones are seen throughout the province as marker posts. 'Macadamized' roads were introduced to Portugal from 1824 onwards, initially at Campo Grande in Lisbon, but even today they have not replaced the *calçada* system. Pink 'granites' of similar syenitic form are known from the eastern Algarve.

Similar but smaller blocks known as *calçadas* are produced from softer rocks, such as the sandstones near Odiáxere and Guia. The most common size of such stones is a 6 cm cube. These *calçadas* are again hewn by hand using a hammer and chisel. The main use of the softer stones is in paving and for ornamental work in gardens. Stones of different colours are used in this work, the most common being known respectively as *calcáreo branco* (white, and very common in the Algarve), and *calcáreo preto* (black, from central Portugal). Both are bedded on to flexible *areão* (grit) and when tapped down with *a martelo de pá* (hammer) will withstand, for many centuries, large dynamic and static loads without deformation. Small stone blocks were widely used all over the Algarve for the lining of hand-dug water wells. Larger blocks of stone from Monchique are cut by hand with great skill and labour into grinding stones used in the province since the earliest times for olive presses. Until the middle of this century, other large stones were cut into cylinders (*cilindros de pedra*), drawn by animals, and used for compaction of the stone blocks used in roadmaking.

Freshwater

Freshwater systems are rather similar throughout Europe and, in the Algarve, include, besides the artificial reservoirs, *albufeiras*, a number of freshwater lakes in the Vila do Bispo area, as well as the river systems, the estuaries of which were discussed in Chapter 2.

The striking plants of Algarve rivers include the giant grass, *Arundo donax* (Gramineae, Plates 195 and 196), known as *cana*, which is conspicuous particularly along the banks of the Guadiana near Alcoutim, and along some of the small rivers between Arão and the Barragem da Bravura. In the Algarve, the best quality *canas* are grown around Tavira and Silves. *A. donax* is generally held not to be native in Europe but to have been introduced from Asia. Walking sticks and fishing-rods are made from the stems, which also provide a source of cellulose used in rayon-making.

Musical instruments have been made from canes for over 5000 years and are the 'reeds' ('Spanish cane') used in clarinets and organ pipes as well as other woodwind instruments, which have also been made for many centuries and are of a primitive form peculiar to the Algarve. These unusual instruments are used to accompany regional *folclórico* dances.

Under favourable conditions, the reed grows to 8 m in height. It is harvested annually, and at longer intervals for canes of the largest size, the mature canes being cut at the base, then trimmed and sun dried. *Canas* have been used for many years in

Figure 219 *Arundo donax* used as traditional under-roofing in a modern house.

Figure 220 Basket-making with *vime* (willow) from Monchique.

Figure 221 *Vime* baskets.

the traditional Algarve houses as the under-roofing (*encaniçado*) to support the tiles above. In the province, the split stems are still made into mats for drying figs and other produce on the flat roofs of houses as well as for covering crushed ice in the local fishing industry. The poles up to 7 m trimmed length, are used in the harvesting of olives and almonds in the autumn (see Chapter 4). Around Odeleite, baskets made from the split stems of the cane include some types exported to France. Similar baskets are made around Monchique, but here generally from split chestnut saplings common in that area (see above). Others are used for packing fruit, shellfish, and other produce for general household purposes (see Chapter 3) and for 'cane' chairs. The canes are also seen in the small *hortas* and *quintas* in vegetable-growing, for example, as support sticks for climbing beans ('French' and some 'runner' types), *feijão verde*, and for tomatoes, *tomates*, as well as fences, both as living and dried canes.

Two other plants are prominent in the river-channels, being submerged at times of high water-levels, and above water-level in the summer. Such plants, known as rheophytes, are recognizable as a class of plant-form throughout the world. Typically, they have narrow leaves and pliant branches. The pre-eminent ones in the Algarve are the oleander, *Nerium oleander* (Apocynaceae) and the tamarisk, *Tamarix africana* (Tamaricaceae). Especially fine examples of them growing together can be seen in the riverbed above the spa town of Alte. Oleander (*loendro*) is grown as an ornamental in parks, gardens, and by the roadside. In 1790, it was recorded that it was grown commercially in the Algarve, the ashes being used medicinally in the thermal baths at Alcoutim and elsewhere.

Figure 222   Rheophytes near Alte, *Tamarix africana* (in flower) and oleander.

In the *Sotavento*, the vegetation of the upper reaches of the rivers has the typical *mato* species, *Cistus ladanifer*, *Daphne gnidium*, *Phlomis purpurea*, *Lavandula viridis*, and *Asphodelus ramosus*, as well as the oleander. Lower down, oleander is found with *Cistus monspeliensis*, *Securinega tinctoria* (Euphorbiaceae), *Scirpus holoschoenus* (Cyperaceae), *Juncus effusus* (Juncaceae), *Lythrum portula* (Lythraceae), as well as the

Figure 223   The lower Aljezur river.

Figure 224   *Alytes cisternasii.* (From Crespo and Oliveira 1989.)

tamarisk and the cane. The lower reaches there and at Aljezur for example, have stands of ash *Fraxinus angustifolia* (Oleaceae), alder *Alnus glutinosa* (Betulaceae), and *Salix salviifolia* (Salicaceae), a willow (its withies used in furniture-making), with the *Scirpus* sp., *Lythrum salicaria*, and *Equisetum ramosissimum* as well as thickets of the bramble *Rubus ulmifolius*, and dog-rose *Rosa canina* (Plate 197) which flowers from the end of April.

Along these rivers are to be found the newt *Triturus boscai*, a species restricted to
Portugal and westernmost Spain, whilst in the lower reaches are the toads *Pelobates
cultripes* and *Alytes cisternasii*, which species is restricted to the south-west of the
Iberian peninsula. *Pelobates* is very large and is found all over wet areas of the
Algarve. The *Sotavento* rivers disgorge into estuaries where the tamarisk, cane, the
reed, *Phragmites australis* (Gramineae), and the rushes, *Juncus acutus* and
J. maritimus can be seen. Streams and marshy areas near them in *Barlavento* are
colourful in spring with the yellow flowers of *Iris pseudacorus* (Plate 198), *Baldellia
ranunculoides* (Alismataceae, Plate 199), and the delicate three-parted white flowers
of *Alisma plantago-aquatica* (Plate 200). In such places, the woody composite
Dittrichia (Inula) viscosa is to be found.

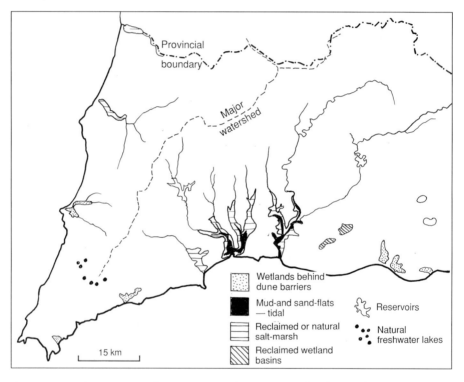

Figure 225   The western Algarve: past and present wetlands. (Redrawn from
Pullan 1988*c*.)

The seasonal lakes near Vila do Bispo are on Pliocene sediments at 130 – 140 m.
They contain water after the winter rains, and their size and duration varies from year
to year with precipitation and evaporation. Lagôa Funda lying to the west of the road
from Vila do Bispo to Aljezur, can have an area of up to 5 ha in some years. Although
the lakes are only temporary, waterlilies (Nymphaeaceae) more commonly seen in the
province's slower-running rivers, have been recorded from them, both the white-
flowered *Nymphaea alba* and the yellow-flowered *Nuphar luteum*. The latter is still to

Figure 226 Plants of wetlands, *as zonas húmidas*. 1. *Alisma plantago-aquatica*;
2. *Lythrum portula*; 3. *Hippuris vulgaris*; 4. *Cotula coronopifolia*; 5. *Scilla odorata*.

be found in flowing rivers east of Silves for example. Amongst the tussocky grasses
and rushes at Lagôa Funda are very tall specimens of *Narcissus bulbocodium* (Plate
201 and 202), some growing to 40 cm. More expected aquatics include *Alisma
plantago-aquatica* and *Baldellia ranunculoides*. In disturbed areas there is much of the
introduced *Cotula coronopifolia* (Plates 203 and 204), native in South Africa; it is also
prominent on the A Rocha peninsula.

The absence of perennial hydrophytes such as waterlilies from Lagôa Funda now
suggests that open water is not permanent there and that, much of the year, the area is
swampy pasture. Around it, field-classes from Oxford University have found a
distinctive transition from the dry *Cistus crispus* and *C. ladanifer* degraded *matos* with
Linaria amethystea (Plate 205), through a band of *Dittrichia viscosa* to one marked by
the presence of the blue- (or occasionally white-) flowered *Scilla odorata* flowering in
March. There are tussocks of the rush *Juncus effusus* and, in deeper water, the mare's-
tail *Hippuris vulgaris* (Hippuridaceae), with *Eleocharis palustris* (Cyperaceae),

Littorella uniflora (Plantaginaceae), and water crowfoots in the *Ranunculus aquatilis* group — *R. peltatus* (Plates 206 and 207), with large flowers, and the minute-flowered *R. omiophyllus* occurring on mud around the edge of the lake — and the starwort *Callitriche stagnalis*, with a species of *Myriophyllum* in the deepest water. By early May, the marshy areas are dominated by magnificent clumps of orchids in the genus *Serapias*, species of which readily hybridize. The large-flowered one is *S. cordigera* (Plate 208), that with smaller flowers *S. lingua* which, in turn, has larger flowers than the common *S. parviflora* of the *matos*. Also in flower at this time are *Myosotis debilis* (Boraginaceae), a forget-me-not, *Scirpus cernuus* (Cyperaceae), *Lythrum junceum* (Lythraceae), and *Silene laeta* (Caryophyllaceae).

7

Streets, parks, and gardens

In the urbanized areas, certain indigenous trees have been maintained, as well as exotics planted. Some of these trees are of remarkably large dimensions, and the Portuguese Government has given them the status of national monuments.

At São Brás de Alportel, on the left-hand side of the road leading to Barranco do Velho, but now being surrounded by housing development, is the largest *Quercus ilex* ssp. *ballota* (Plate 209) in the province. Declared a monument in 1942, this oak has a crown diameter of 25 m and a bole diameter at chest height of 4.2 m. A tree with a crown diameter of 29 m but with a smaller bole, is recorded at São Marcos da Serra on the plain of the Odelouca stream (the *Baião* holm-oak), but it is in the Monchique area that trees of great dimensions are mostly found. Two Norfolk Island pines, *Araucaria heterophylla* (Plate 210), a species from Norfolk Island in the Pacific and usually grown as a seedling in conservatories in northern Europe, can be seen there. The larger of these is 35 m high with a diameter at breast height of 3.9 m, the other tree only a little smaller. One is clearly visible from the square in the middle of the town, whilst from the other side of the square, but facing west and on a steep hill, what is possibly the largest *Magnolia grandiflora* (Plate 211) in Europe can be seen. The tree is native in the south-eastern United States and has glossy dark green leaves and creamy-white flowers up to 25 cm in diameter. The Monchique tree, which is in the grounds of the derelict Convento de Nossa Senhora do Desterro, is at least 200 years old and over 26 m tall. It has a trunk girth of 5.3 m and a crown diameter of 25 m, though recent wind damage has nearly severed one of its major limbs.

Between Monchique and Alferce are some very large strawberry trees, *Arbutus unedo*. Of the commonly grown acacias or wattles, which include *Acacia retinodes* (Plate 212), 2 km north of Monchique, there are tall specimens of *Acacia melanoxylon* (*Acácia Austrália*) and also *Eucalyptus globulus*, both native in Australia, and, in nearby Barranco dos Pisões, there is a London Plane, *Platanus x hispanica*, which is 35 m tall and with a girth of 4.7 m. This tree is believed to be a hybrid between *P. orientalis* of south-eastern Europe to northern Iran, and the north American *P. occidentalis*, raised in the Botanic Garden at the University of Oxford from seed received from Montpellier in the seventeenth century. It is much planted in towns, as it resists pollution because of its rapidly flaking bark.

Most of the oldest olives in the province are those in the south-east near Tavira, one of the most remarkable, as was noted in Chapter 4, being in the tourist complex at Pedras d'El-Rei (Plate 213). This tree has a split and hollow bole and with a girth of over 10 m; it is said to be 2000 years old. The *Oliveira do Purgatório*, near Messines on the *Barrocal*, is an olive tree with a girth of 6.96 m. Of carobs, there are recorded specimens to 30 m crown diameter and 6–8 m girth. Some fine examples of such trees are found around Loulé, and between Olhão and Tavira. At Vale d'El Rei, near Lagôa, there is an exceptional specimen of the 'umbrella pine' (*pinheiro manso*, *Pinus pinea*).

Commonly planted in towns, notably in Faro, in which city there is a very large example in Alameda João de Deus, and in Lagos, are avenues of the Canary date-palm *Phoenix canariensis* (Plates 214 and 215), much grown in southern Europe for ornament, but restricted in the wild to the Canary Islands. *Phoenix canariensis* is easily grown from seed, although great care must be taken in transplanting to avoid damage to the tap-root. It differs from the cultivated date palm *Phoenix dactylifera* — which is little grown in the Algarve (but see Frontispiece) — in that it has a single very stout stem with fruits up to 2.3 cm long. These fruits, ready in autumn, are nutritious and taste very like those of the date palm, which often has several stems, these being more slender, and has longer fruits, up to 7.5 cm. It is grown in plantations near Elche in south-eastern Spain, an area with a microclimate which suits this plant. It is unknown in the wild, its closest ally being *P. theophrasti*, a recently discovered, and now rare, species found at a few sites in Crete and Turkey.

In some towns such as Lagôa and Lagos, are shady groves of sweet-smelling citrus. Street-trees generally are hybrid poplars, the timber of which is used for punnets for fruit such as strawberries, although another application in the Algarve is for toothpicks, as the wood is scentless. Also found in street plantings, notably in Portimão, is the 'nettle-tree' or hackberry, *Celtis australis* (Plates 216 and 217), flowering from the end of April. It is a native of southern Europe, and is widely planted for shade. The timber is valuable and the fruit is edible. Also in Portimão, one

Figure 227   Old specimens of *Phoenix canariensis*, Portimão.

Figure 228 Roadside poplars.

square is surrounded by the tree-privet *Ligustrum lucidum* (Plate 218), a Chinese species widely cultivated in Europe and Japan. It has typical privet flowers from June, followed by trusses of purple-black globose fruits, which persist until March. Also in Portimão and elsewhere, for example in the castle at Silves, are fine specimens of the Persian lilac or bead-tree *Melia azedarach* (Meliaceae, Plates 219 and 220). Flowering from the middle of April, this is an Indian species cultivated since the time of the Persian Empire, and it is now widely naturalized throughout warm parts of the world. Its natural distribution extends to Australia, where it is the timber tree known as white cedar. The fruits contain a hard endocarp dimpled at each end, 'pre-adapting' them to be beads for rosaries, for which they are still used. The plant is an effective insect repellent, dried leaves put into books keeping them free of insect attack. The fruits persist on the branches until the following year and are toxic to humans and pigs.

Other street-trees (Plates 221–5) include the magnificent azure blue-flowered jacarandas, flowering on otherwise almost bare branches from April onwards, with the display lasting until October. These are cultivated forms of *Jacaranda mimosifolia* (Bignoniaceae) introduced from north-west Argentina in the early nineteenth century. Also much planted is the so-called 'Judas Tree' *Cercis siliquastrum* (*olaia*, Leguminosae), to be seen flowering in spring along country roads as well as in towns. Many are planted on the road from Portimão to Caldas de Monchique and there are other good specimens in the courtyard of the Parish Church (*Igreja Matriz*) of Portimão. Readily raised from seed, *C. siliquastrum* is a deciduous tree with camel-foot-shaped leaves and beautiful mauve flowers produced on the main branches and even the trunk. It is native to the Mediterranean region (but not to Portugal) and a confused legend has it that Judas Iscariot hanged himself on one. This is probably not so, as *Cercis* is *Arbor Judaeae*, namely 'Judaea tree', as it was commonly cultivated around Jerusalem, and not *Arbor Judae*, which in the Mediterranean is associated with a fig. The story has become even more confused in Great Britain where the gibbet is associated with the foul-smelling elder-tree *Sambucus nigra* (Caprifoliaceae), known in Kent as the Judas tree, the edible fungus (*Auricularia auricula-judae*) on it being called Jew's ear.

The tree seen most often in churchyards is the Italian cypress, the pencil-like *Cupressus sempervirens* (Plate 226), found throughout the Mediterranean region. This is a cultivated form of the wild forma *horizontalis* native to southern Europe and south-west Asia. The Egyptians used the timber for their sarcophagi, and the Greeks for statues of their gods. Traditionally, the branches were used as an infusion in a foot-bath to combat smelly feet, and it is still a constituent of certain scented soaps. The tree grows to a height of 25 m and has large pear-shaped cones. It has very low water requirements, although it is not an ideal tree for planting as an isolated specimen in exposed places, since the root system is not able to prevent windthrow.

The pepper-tree *Schinus molle* (Anacardiaceae, Plate 227), from the Peruvian Andes, is much used as a shade-tree. In South America it has a number of medicinal uses and the pink fruits have been used to adulterate pepper. Exudates from the tree have been chewed (American mastic) and in Uruguay they have been used for fertility control. However, it harbours the pestilential black scale of citrus (*Saissetia oleae, cochinilha preta*). Differing in its smaller, bright-red fruits, the allied *Schinus terebinthifolius* is naturalized in Florida, where it is a cause of dermatitis and of respiratory problems, like so many Anacardiaceae, which family includes the poison ivies and poison oaks of the United States. None the less, it provides good bee-forage in the form of nectar. *Schinus molle* grows in the Algarve, even on limy soils, and is

Figure 229   Roadside cypresses.

locally naturalized. It develops rapidly but in spite of an extensive root-system, it does not anchor well, and tends to lean so that it rarely forms an elegant specimen tree.

Many woody plants of the rural landscape are cultivated in towns for ornament, as are many succulents (Plates 228–33), such as the yellow-flowered *Aeonium arboreum* (Crassulaceae) from Morocco, which is seen in the Algarve in two forms, one with pale-green leaves, the other (cv. Atropurpureum) with deep red leaves, *Lampranthus* species, and cacti, as well as the statuesque agaves and yuccas. There are some 40 species of *Yucca* (Agavaceae), all of them native in the warmer parts of north America, and all having stout stems, stiff, sword-like leaves, and panicles of off-white flowers, which in the Algarve are seen in September. Propagation of yuccas is easy from broken-off limbs, also from pieces of the root system and from offsets developing underground from the parent plant. Although seed is produced in gardens, the large seed capsules of *Y. aloifolia* for example, with viable seed allowing the plant to escape, most species in the wild depend on yucca moths for pollination and successful seed-set.

All those *Yucca* species found east of the Rockies are visited by *Tegeticula yuccasella*, which is active by night, by day resting in flowers, which are of a colour similar to the moths. The flowers are most strongly scented at night and nectar is sometimes secreted at the base of the ovary, although the moths do not take it; it may possibly serve to attract other insects away from the stigma. The female yucca moth climbs up a stamen and bends her head over the anther, steadied by her uncoiling tongue, the pollen then being scraped together in a ball under her head by her maxillary palps. Up to four stamens are processed before the moth flies to another flower, where the ovary is inspected. If the ovary is of the right age, and does not already have eggs in it, the moth usually lays one egg in each cell and, after laying each one, deposits some pollen into the tube formed by the stigmas. As unpollinated flowers are soon dropped, the deposition of the pollen ensures that there will be a continuing supply of food for the larvae provided by the growth of fertilized ovules near them. Other ovules develop normally, and the larvae emerge to pupate in the ground when the seeds are ripe. The adults emerge over a period of up to three years, so that if the yuccas nearby do not flower every year, then at least some moths will survive to reproduce. West of the Rockies, the yucca moths are more particular as to the species visited. False yucca moths, *Prodoxus* spp., also breed in the ovaries but do not pollinate, thereby relying on the true yucca moths for their survival, a system very much like that of the parasitic fig wasps described in Chapter 4.

Of those species seen in the Algarve, *Y. aloifolia* ('Spanish bayonet'), introduced into Europe by 1605, has edible fruits and its leaves have been used as a source of fibre. Also grown are the two larger species, the east North American *Y. filamentosa*, introduced by 1675, and *Y. gloriosa*, the first known species introduced by 1550, which can easily be distinguished from one another in that the first has leaves 2–5 cm wide, those of the second exceeding 7 cm.

Hedge-plants include *Myoporum tenuifolium*, described on p. 88, and sometimes mixed with, or replaced by, *Pittosporum tobira* (Plate 234), a small tree with heavily scented flowers introduced from the Far East to Europe in the early nineteenth century. *Myoporum tenuifolium* is an important windbreak in Barlavento, while in Sotavento the Australian *Hakea salicifolia* (Proteaceae) is widely planted for the same reason.

Other plants used in ornamental hedges are cultivars of the hibiscus, *Hibiscus rosa-sinensis* (Plates 235 and 236), a plant grown for so long in Asia that its origins are now obscure. There is a superb example nearly 3 m tall and over 100 m long at Cabeço do Estevão, near Alcalar (Plate 237), where it flowers from June until November. *Hibiscus* flowers are also found in 'double' forms in a range of colours, although these are rarely as vigorous as the 'singles'. Another, less attractive, hedge plant is *Lantana camara* (Verbenaceae, Plates 238 and 239), native in tropical America, but widely introduced elsewhere, and naturalized in the Azores. *Lantana* is often a serious weed in many tropical countries, where it takes over native vegetation. Its juicy fruits are dispersed by birds, and its introduction or cultivation in many countries is prohibited. Sterile forms propagated vegetatively are available, and there are a host of forms with flowers of different colours. Many of them grown in the Algarve are sterile, which may explain why it is not as invasive as in neighbouring Andalucia.

Numbers of these trees and shrubs are grown in private gardens, but Algarvians have time only for productive trees, any ornamentals grown being largely herbaceous. An exception is the shrubby *Spiraea cantoniensis* (Plate 240), which has white flowers and irregularly lobed single leaves. It was introduced from the Far East early in the last century and is to be seen as a cutflower in markets. In municipal plantings, hotel and private gardens, there is the usual Mediterranean riot of exotics including cultivars of *Canna indica* (Plate 241), originally from tropical America, one form being the 'Queensland arrowroot', neither from Queensland originally, nor a true arrowroot, and pelargoniums, usually hybrids derived from southern African species. From southern Africa also are the osteospermums, sometimes incorrectly known as dimorphothecas, the most familiar being *O. ecklonis* (Plate 242) with white flower-heads, the central flowers being blue. Forms of the native *Chrysanthemum coronarium* are grown, as are the Canarian *Argyranthemum (Chrysanthemum) frutescens*, with white heads of flowers in April and the shrubby and vivid blue-flowered *Echium* species, also mostly from the Canary Islands.

The 'chrysanthemums' of gardens are forms of *Dendranthema × grandiflorum*, a complex hybrid group raised in China from *D. indicum* and other species. In Portuguese gardens, many are cut on All Saints Day (1 November), when they are taken to the local cemetery, All Souls Day being 2 November. Of other plants often seen, *Acanthus mollis* ('bear's breeches') is a native species flowering (also on roadsides) from the end of April, gazanias (Plates 243 and 244), largely *G. rigens*, and *Arctotis* hybrids, again derived from southern African plants, thrive in the Algarve. They are of many different colours, but all have brilliant flowerheads, opening only to the sun. They flower over a long period from April to November, and are propagated by division when dormant. Familiar as in northern Europe, are nasturtiums (*Tropaeolum majus*) and the gaudy Californian poppies (*Eschscholzia californica*, Papaveraceae, Plate 245), which were introduced from California to Europe before 1800. Also often found in gardens and perhaps cultivated for many years, is the 'double' opium poppy, *Papaver somniferum* 'Paeoniiflorum' (Plate 246), with its blowsy over-heavy flowers.

A familiar cottage-garden plant grown in northern Europe is the highly scented Madonna lily *Lilium candidum* (Plate 247), which in the Algarve produces its white flowers at the end of April on metre-high stalks. They are pollinated by hawk-moths.

Possibly originating from south-west Asia, this lily is naturalized in parts of Europe, and has been grown for its scent, some 500 kg of flowers yielding 300 g of essence. The plant is depicted in Cretan frescos 5000 years old and was possibly the Rose of Sharon of the Bible. It has been cultivated since at least 1500 BC and, in the Christian era, has become associated with the Virgin Mary, who is often depicted in the works of Botticelli and Titian holding this lily in her hand.

Cutflowers in markets are often mixtures of *matos* plants with exotics, some of which are almost naturalized, particularly the arum lily *Zantedeschia aethiopica*, originally from southern Africa. Freesias, flowering in March, are largely hybrids raised in earnest since the end of the last century and referred to as *F.* x *hybrida*, derived from southern Africa species, with the very fragrant greenish-yellow *F. refracta*, the strongly-scented white *F. alba*, and the pink forms of *F. corymbosa* all being involved. Nowadays too, there are 'double' forms, tetraploids, etc. The ability to smell them varies from person to person, and some find it difficult to detect any scent at all. Heavily-scented tall carnations are grown commercially in *estufas*, but often in gardens too.

Plants from many parts of the world can be grown in the Algarve, but the success of those from 'Mediterranean' climates elsewhere is manifest in the number of Australian and southern African species to be found naturalized as well as cultivation. Some so far unmentioned include the bottle-brushes, *Callistemon* spp. (Myrtaceae, Plate 248) from Australia, notable in that the tuft of leaves at the top of the infloresence continues growth after the flowers fade. In the wild the flowers are pollinated by birds. Species and hybrids of the tropical American genus *Brugmansia* (Plate 249) with their large dropping 'angel's trumpets' ('daturas' but true *Datura* species like the common *D. stramonium*, an annual species of roadsides and wasteland throughout Europe, have erect flowers), are also common. *D. stramonium* was grown towards the end of the eighteenth century around Monchique, where its medicinal properties were recognized. In the same place and for the same reason, peonies and various species of *Digitalis*, *Solanum*, *Nepeta*, and *Veronica*, were grown commercially. Possibly originating in north-east Africa, the castor-oil plant (*Ricinus communis*, *ricino*, *Palma Christi*, Plates 250 and 251) is frequently seen in gardens and forms of it are now grown both for their scarlet ornamental leaves and for their repellent action to mosquitoes and flies. It used to be grown commercially in the province, as it is a valuable source of 'castor-oil' (*óleo de ricino*), which was used for lubricating machinery in the First World War when it was considered the best for aero-engines, and as an illuminating oil and for many other industrial purposes. Medical uses include its application in X-ray imaging (as *leite de ricino*).

The list of cultivated ornamentals in the Algarve is very large as so many plants are hardy here, so that the scarlet-flowered *Tecomaria capensis* (Bignoniaceae, Plate 252) from southern tropical Africa, flowering from April to the end of October, shrubby milkworts *Polygala* from southern Africa, and nightshades *Solanum* from South America, rubber plants *Ficus elastica* (Plate 253) from Asia, and cycads from Africa, will be seen growing together. Some of the many other cultivated ornamentals are shown in Plates 254–69. The general effect is of lush growth, but roses are an exception. Common in the country, though, is a vigorous climbing Bourbon rose with neat clusters of double-white flowers, opening from crimson buds. This is 'Félicité et Perpétue' (Plate 270), raised in 1827 by the gardener to the Duc d'Orléans. It is widely grown in Europe, thriving in most soils and positions, including shade. Because of the almost continuous growing season, hybrid tea roses are generally not very vigorous for long periods, though robust cultivars similar to floribundas, notably the pink-flowered 'Queen Elizabeth' (Plate 271), named in honour of the British Queen Mother, when Queen of England, are successfully grown.

Other climbers frequently seen include bougainvilleas (Plate 272), commonly reds, but also orange and white, the colour of the 'flowers' due to the persistent bracts surrounding the minute true flowers; these climbers flower for some nine months starting in April and are armed with stout spines, though prone to attack by rats, which seem to find excellent meals in the tough stems. *Wisteria* flowers for only a few weeks, and most notable is the beautiful *W. floribunda* (Plate 273), which has inflorescences of scented flowers to over 100 cm in length, and leaves with 13–19 leaflets compared with the less desirable but more often seen *W. sinensis*, which has much shorter inflorescences of scentless flowers and leaves with only 11 leaflets. It is interesting that the first, from Japan, climbs clockwise, the second, from China, anticlockwise. Both have poisonous leaves and seeds like their ally, the laburnum. Even more exotic is the rampant *Solandra maxima* (Solanaceae, Plate 274), the chalice vine from Mexico, introduced to Europe in 1781 and flowering from May. The showy flowers are up to 20 cm long and are night-scented. They burst open, one day a brilliant yellow, but the following one, have turned to a dirty brown, and are old gold by the third. In Mexico, it and its allies were the source of sacred hallucinogens.

Lawns and golf-courses

Fine lawns of the type for which Britain is noted are rarely seen in the province. The major grasses found mixed in British lawns are species such as common bent (*Agrostis capillaris*), Chewing's fescue (*Festuca rubra* ssp. *commutata*), and perennial rye-grass (*Lolium perenne*). Such mixtures are not tough enough to withstand the intense heat and drought between May and September in the Algarve. Moreover, creeping perennial weeds, especially the sedge *Cyperus rotundus* ('the world's worst weed'), grow rapidly and are efficient competitors of the grasses. Turfing, a popular method of establishing ornamental lawns in Britain, is rarely practised in the province, partly because the required field grasses are hardly ever available, but also because transplanting turves is a near-hopeless task since the roots have insufficient depth to tolerate the very dry conditions. Coarse spreading grasses are more successful, if watered in dry periods and, since most of these root readily from the nodes, individual pieces can be planted at any time from early spring to early autumn if well watered in. Of the grasses commonly grown, those most often seen are 'Kikuyu' (*Pennisetum clandestimum*) from tropical Africa and the native 'Bermuda' (*Cynodon dactylon*), both not as coarse as the widely planted alternatives, the American 'St. Augustine' (*Stenotaphrum secundatum*) and 'Centipede' (*Eremochloa ophiuroides*) from south-east Asia. All these are very tough, quickly forming dense mats, as they spread by underground shoots. Maintenance can be reduced if 'pop-up' sprinklers or other distribution systems for water are incorporated throughout the lawn.

On golf-courses in the Algarve fairly coarse creeping grasses or mixtures including tough annual rye grasses such as *Lolium multiflorum* are used for the fairways, usually

209 *Quercus ilex* ssp. *ballota* at São Brás de Alportel

210 *Araucaria heterophylla* at Monchique

211 *Magnolia grandiflora* above Monchique

212 *Acacia retinodes*

213 The olive at Pedras d'El Rei

214 *Phoenix canariensis* at Faro

215 Fruits of *Phoenix canariensis*

216 *Celtis australis*, Portimão

217 Fruits of *Celtis australis*

218 *Ligustrum lucidum*, Portimão

219 *Melia azedarach*, Silves Castle

221 *Robinia pseudoacacia*

222 Jacarandas at Lagoa

225 *Cercis siliquastrum*

220 *Melia azedarach*, fruits

223 Jacaranda flowers

224 Jacaranda fruits

226   Cypresses in the graveyard at Alte

227   *Schinus molle*

228   *Aeonium arboreum*

229   *Agave americana* in fruit

230   *Agave americana* cv. Marginata

231   *Portulacaria afra*

232   *Lampranthus* cv.

234   *Pittosporum tobira*

233   *Lampranthus* cv.

235   *Hibiscus rosa-sinensis*

236   Red-flowered *Hibiscus rosa-sinensis*

237   *Hibiscus rosa-sinensis* hedge

238   *Lantana camara* hedge

239   *Lantana camara* flowers

240   *Spiraea cantoniensis*

241   *Canna indica*

242   *Osteospermum ecklonis*

243   *Gazania* cv.

244   *Gazania* cv.

245   *Eschscholzia californica*

246   *Papaver somniferum* 'Paeoniiflorum'

247   *Lilium candidum*

249   *Brugmansia* hybrid

250   *Ricinus communis* flowers

248   *Callistemon* sp.

251   *Ricinus communis* fruits

252   *Tecomaria capensis*

253   *Ficus elastica*

254   Pink-flowered oleander

258   *Dombeya × cayeuxii* (Sterculiaceae) hybrid of *D. burgessiae* from south-east Africa and *D. wallichii* from Madagascar

255   White-flowered oleander

259   *Abutilon* cv. (Malvaceae)

256   White oleander flowers

261   *Senecio angulatus* flowers

257   *Cordyline australis* (Agavaceae) from New Zealand

260   *Senecio angulatus* from southern Africa

262 *Euphorbia pulcherrima*, poinsettia, from Mexico

264 *Plumbago auriculata* from southern Africa

266 *Cortaderia selloana* (pampas grass) from Argentina

263 Variegated lemon, fruits and leaves

267 *Hemerocallis* (daylily) hybrids

268 Ivy-leaved pelargoniums originally from southern Africa

269 *Ornithogalum thyrsoides* (chincherinchee) from southern Africa

265 *Impatiens sultani* cv. (Busy Lizzie) from tropical East Africa

270 Rose 'Félicité et Perpétue'

271 Rose 'Queen Elizabeth'

272 Bougainvillea

273 *Wisteria floribunda* at Portimão

274 *Solandra maxima*

275 *Antirrhinum majus* in Portimão backstreets

276 *Hyoscyamus albus*

277 *Nicotiana glauca*

278 *Nicotiana glauca* flowers

279 *Umbilicus rupestris* on roof tiles

Figure 230   Golf-course at Penina.

with finer grass mixtures for the greens. The first golf-course in the province was made near Praia da Rocha in the 1920s, but that had sand-greens. With the opening of the airport at Faro, improved road communications with Lisbon and the expansion of popular tourism since 1960, golf-course construction has proceeded rapidly to meet the increasing demand. Already, there are 18 courses, of which five are still being constructed (1991). Those well established are, from the west, Parque da Floresta, Palmares, Penina, Alto Club (Alvôr), Vale de Milho (Praia de Centianes), Pine Cliffs, Vilamoura I, II and III, Vale do Lobo, Quinta do Lago, and São Lorenço. Subsequently modified, but originally created from rice-fields under the direction of the three-times winner of the British Open Golf Championship, Sir Henry Cotton, the first major one was the Penina course opened in 1966.

Walls and roofs

The sides of the streets themselves, when not too assiduously weeded, have an array of arable weeds, typical of the countryside (Plate 275). Even walls and roofs have some interesting plants. On the castles at Silves and Castro Marim, as well as on the city walls of Faro, are healthy specimens of the henbane *Hyoscyamus albus* (Plate 276) which, like *H. niger* (differing in its yellow, purple-veined, as opposed to pale-yellow flowers), is a source of alkaloidal drugs used as hypnotics and narcotics. The old city walls of Faro have a range of vascular plants including *Parietaria judaica* (*P. punctata*, Urticaceae), *Mercurialis annua* (Euphorbiaceae), *Centranthus calcitrapae* and *C. ruber* (Valerianaceae), a European plant naturalized in California, but

Figure 231 Wall flora, Faro.

most spectacularly perhaps, the woody tobacco relation *Nicotiana glauca* (Solanaceae, Plates 277 and 278), introduced from South America early in the last century. Native plants found on house-roofs in such places include the succulent *Umbilicus rupestris* (Plate 279) all over the low-pitched roofs of the older houses as at Castro Marim.

8

The Algarve, past and future

Having surveyed the plant-life, both wild and cultivated, of the Algarve, we must now show how it fits in with the ever-changing physical environment of the region, particularly examining the changes wrought by human activity through time.

Physical change

The Algarve lies in a tectonic zone with a large number of faults. Off southern continental Portugal, there is an area of the seabed where the cold and brittle upper crust forms the lithospheric shell which overlays an area known as the asthenosphere, this being made up of a series of rigid units (tectonic plates), each of which has a separate motion relative to the other plates. Earthquakes (*terramotos*) result from the release of the stresses caused by these rigidities and motions. The epicentres of these earth shocks are mostly between the Gulf of Cadiz, an area of the seabed which is much fractured, and a large area south-west of Cape St. Vincent, together with a further cluster around Lisbon.

Between 1755 and 1973, a total of 1039 earthquakes was recorded in the country, the areas of continental Portugal most affected being Douro-Minho, Estremadura and parts of Alentejo and the Algarve. Around Lisbon itself, there have been 117 significant tremors since 1147, the largest of them in 1531, 1597, and 1755. On the Portuguese mainland as a whole, the effects of a series of major earthquakes were felt on 24 August 1356, 26 January 1531, then a number in the eighteenth century — 6 March 1719 and 27 December 1722, crowned by the disaster of 1 November 1755. The last was of sufficient force to destroy much of Lisbon and the Algarve; it was followed by a long series of further tremors which finally ceased on 20 August 1756. In the Algarve, there was a tidal wave almost 20 m high and sand was deposited in harbours and river-mouths, as at Alvôr and Aljezur, where the lower reaches of the Aljezur river were irrevocably salinated. Even at Monchique the tremors had serious effects, and the fortifications of Lagos were destroyed; they were still in ruins in 1908. The church and many large buildings in Portimão were badly damaged and the old Moorish walls were later dismantled to provide stone for rebuilding the church. At Sagres, the neglected School of Navigation was completely destroyed.

Another earthquake on 11 November 1858 devastated the city of Setúbal, close to Lisbon. Two major earthquakes in 1960 destroyed Agadir in neighbouring Morocco, the seabed between Morocco and Portugal forming part of the same tectonic plate system, and killed more than 20 000 people. Minor earthquakes are frequent in the Algarve; one in 1969 damaged several dozen houses at Villa do Bispo. There were others centred round Loulé in the 1970s, whilst several small tremors have been felt in the 1980s, and one in 1990, in the province. The effect of all these earthquakes on the flora and fauna of the Algarve is not easy to evaluate. Clearly, there has been much physical destruction on land and associated large fires have destroyed much, while the coastline has been greatly altered over time.

Ecological change

Plant fossils of the Upper Cretaceous in Provence have only one readily identifiable 'Mediterranean' genus — *Laurus*, though the progenitors of Rhamnaceae, Legum-inosae, Liliaceae, and *Quercus* may have been present, but in Westphalia an alleged *Nerium* species has been recorded from this era. Half of the flora of a very rich early Eocene deposit in northern France is not reconcilable with living genera; such floras are considered to be referable to tropical (Indomalesian) assemblages so that 'Mediterranean' elements would hardly be expected from them. The London Clay Flora of that period has a conifer, considered by some to be the antecedent of *Tetraclinis articulata*, now restricted to north Africa, south-east Spain, and Malta. The early Oligocene in the Loire has a leaf flora containing a form of *Pistacia lentiscus* but, by the end of the Oligocene, at Aix-en-Provence 21 of the 89 recorded genera of fossils are represented in the modern flora there. Of the 231 species, 17 have been referred to modern species including *Ceratonia siliqua*, *Cercis siliquastrum*, *Ficus carica*, *Juniperus* cf. *phoenicea*, *Nerium oleander*, *Pistacia lentiscus*, *Quercus ilex–coccifera* types, *Rhus coriaria*, and *Tetraclinis articulata*, many of which have been subsequently reintroduced to the western Mediterranean by humans. By the Pliocene, the pollen record shows gymnosperms such as species of *Picea* (spruce), *Tsuga* ('*Cathaya*'), *Pinus*, Taxodiaceae, and Cupressaceae, with a wide range of angiosperms including *Quercus*, *Carya* (hickory), and Ulmaceae. At Rio Maior in the Tagus Basin of Portugal, deposits of this period show the presence of *Cistus*, Cupressaceae, *Helianthemum*, *Myrtus*, Rhamnaceae, and *Vitis*. By the end of the early Pliocene, many 'exotic' taxa had become very rare or had disappeared, for example Taxodiaceae, apparently during a period of decreasing rainfall.

Pollen studies reveal marked fluctuations of two main vegetation types between the late Pliocene and early Pleistocene: 'open' communities of *Pinus* and Cupressaceae with herbs, perhaps also with *Ceratonia*, *Olea*, and *Pistacia* in some areas on the one hand, and closed forest communities rich in deciduous species no longer present in the region — *Carya*, *Parrotia persica*, and *Pterocarya* (now Asiatic or north American) for example — on the other. The last million years since the early Pleistocene have had rather regular oscillations of vegetation type associated with northern glaciations, but evidence from the late Pleistocene is sparse and there is none from southern Portugal. There is no evidence of Mediterranean-type vegetation in the western Mediterranean at the height of the last glaciation (160 000–55 000 years ago), though it appears from Provence that there may have been refugia for it during the most severe periods. Late-glacial deposits have high frequencies of *Juniperus* pollen, followed by *Pinus* which declined as oaks moved in, *Juniperus* dropping out only as *Quercus ilex* became dominant, about 7500 years ago.

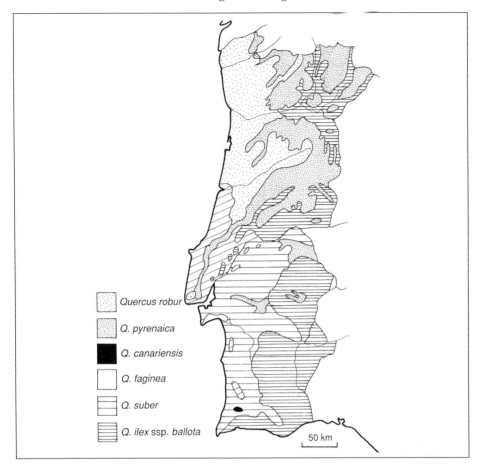

Figure 232 Supposed original distribution of oaks (*Quercus* spp.) in Portugal. *A distribuição das Quercus espontâneas de Portugal.* (Redrawn from Lautensach 1988, p. 597.)

Because of the greatly modified landscape of the Algarve and the paucity of suitable sites for preservation of the pollen record giving an indication of the plants formerly growing there, it is very hard to reconstruct the original vegetation cover of the area. That the distribution of the cultigens olive and carob should have been used to define part of the area floristically has emphasized this difficulty. More recently, the focus of investigations has been the native species of oaks in the original vegetation. This shows the Algarve with cork oak dominating the acid rocks north-east of the *Barrocal* and *Q. canariensis* in the Monchique area, with the *Barrocal* and the rest of the lowlands as well as the east occupied by *Q. ilex* ssp. *ballota*, but a tentative recon-struction of the coastal region shows the lowland area of the west typified by *Chamaerops humilis*, with *Pinus pinea* to the east.

There is evidence of human use of fire in the Mediterranean as far back as the mid-Pleistocene, that is to say, about 500 000 years. However, the earliest definite evidence of human occupation in the Algarve is from Mirouço, north of Vila do Bispo, and this is some 170 000 years old. There are Mesolithic (10 000–3500 BC) indications at Castelejo near Budens, at Cabranosa near Sagres, and at Ludo near Faro. At Cabranosa, there are early Neolithic (5500–3400 BC) remains, notably sherds, and Neolithic-Bronze Age transitional finds have also been made at Monchique. Moreover, at Alcalar, near Portimão, there is a remarkable series of at least seven Neolithic-Bronze Age (2500–1800 BC) structures, *tholoi*, considered by some to be necropoli, although there is no record of any bones found within them, nor indeed is there any clear evidence of religious, military or other purpose. Similar structures have been discovered up the Atlantic coast of Europe, whilst those at Alcalar, which were first (partially) excavated in 1880 and declared National Monuments in 1910, are very similar in dimensions, and in detailed design, to those found in Cornwall, for example at Boleigh, and described there as 'fogous'. In comparison with other prehistoric subterranean structures, fogous present a number of unique features; a combination of subterranean stone cupolas, specific orientation, and one or more side 'creep' passages, sometimes blind, built with lintels and dry-stone walling, with more than one entrance, and often associated with nearby settlements. Little is known of the economy of their builders, who were possibly 'Celts' if the Cornish comparison is valid, and the effect they had on the landscape is unknown. It is possible however that, like their contemporaries in northern Europe, these inhabitants of the Algarve of 4000 years ago exploited the more readily cleared and reasonably fertile lowlands, putting their strong buildings on firm foundations with easy access to nearby ports, rather than occupying the inhospitable shales or the stonier areas of the *Barrocal*.

The uplands of this postglacial period were essentially oak forests, but the vegetation of the lowlands before there were grazing and burning regimes maintaining the modern landscape is more difficult to establish. None the less, trees of various species can be grown in this area, and the suspicion is that the *Litoral* was indeed forested with *Quercus ilex* ssp. *ballota* and *Pinus pinea*, with *Quercus coccifera* and its associates, now making up the typical *matos* landscape of the country, to be found on crags and steep slopes, where these trees could not become established. Except for such rocky or unstable areas, it seems likely that forest, admittedly much dwarfed, would have occurred on the highest mountains in the region and also near the coast, though it is possible that the steep windswept cliffs, both of limestone as at Cape St. Vincent, and shales further north, would, as today, have been covered with *Juniperus phoenicea* and its associates. However, in the Mediterranean as a whole, the most general consequence of man's activities has been the decline or disappearance of deciduous oak forest to the advantage of the sclerophyllous. In this way, *Q. ilex* ssp. *ballota* and *Q. coccifera* have taken over from *Q. faginea* and *Q. canariensis* in the west and, under increasing management involving fire, led to the final dominance of *Q. suber*, which is able to regenerate more readily after burning, producing shoots to 20 cm within a month. Similarly, human activities have increased once more the spread of pines throughout the region.

The corollary of all this is that many plant species have become much more common in historical time, and that the fire-maintained landscape of much of the area is more colourful and species-rich than the forests of oak or pine would have been, though the same species could have been found in different parts of the region. Just as with the species-rich limestone grasslands of the British Isles for example, the effect of humans and their animals has been to bring together, in an anthropogenic assemblage, native plants from different habitats, enriched with exotics introduced deliberately for food, for example the olive, fig, carob, and almond, or accidentally, for example *Oxalis pes-caprae*, in the areas under cultivation, as well as in an uncultivated vegetation type, *matos*. Taken together then, the plants from the forest understorey, and possibly those from coastal cliffs and high altitudes as well as those from the natural clearings in forest, have not evolved together as an assemblage, but are now growing together under the extraordinary regime imposed by humans and their animals. If humans disappeared, would the original forest-cover return? Given the aggressive nature of some of the exotics, still apparently uncontrolled by natural enemies, the native flora would probably *not* return to mirror the original exactly.

Around Algiers, burning leads to degraded *garrigue* dominated by *Chamaerops humilis* and, with greater pressure, erosion and bare rock. It is not clear how long fire has been an important feature of the Algarve's ecology. Certainly, it is known that some species will not tolerate repeated firing, and those areas of the Algarve with frequent burnings tend to have low numbers of plant species. On the other hand, a number of species seem to be 'pre-adapted' to a fire regime, that is, to survive and propagate under such conditions. It is true that many of the features, for example, geophytism, are also associated with the very seasonal climate so that plants with this feature appear to be 'pre-adapted' to the 'new' conditions imposed by human beings. In other regions with a 'Mediterranean climate', for example Australia, certain plant species, such as many *Eucalyptus* and *Acacia* species, are adapted to fire in that their seeds germinate after it or the frequency of flowering is enhanced.

It is now probably impossible to disentangle what was 'originally' adapted to what, because human beings, in imposing a fire regime on the region, have unintentionally initiated a selection experiment favouring those species or those individuals of a particular species which are capable of surviving in such conditions. Moreover, it is difficult to conduct meaningful scientific experiments because the advent of a fire leads to changes in other environmental factors, for example, opening up of the habitat, and therefore, for example, different light and water regimes.

How fire-tolerant might the original forest-lands of the Algarve have been? One suspects that there may well have been fires, but, like other disturbances to the ecosystem which promote diversity locally, they would have been over smaller areas and at a lower frequency than obtains today, except in times of major earthquakes. Nevertheless, such disturbances including those initiated or maintained by browsing and grazing animals, would have given the opportunity for short-lived plants to co-exist with their more robust neighbours. The effect of humans has been to exaggerate this. How many plant species have extended their range into the Mediterranean from further east is unknown, but that many 'Mediterranean' plants can become established in the Algarve — for example *Oxalis pes-caprae* — suggests that this may be a significant factor in the richness of the modern flora.

Fires caused by lightning strikes may be significant as might vulcanism in selecting the species through time. There are 500 000 strikes each day in the world's forests and 50 000 woodland fires are started by lightning each year. Electric storms, with multiple lightning strikes, occur quite often in the Algarve, although localized, and in recent decades they have led to a very serious fire every five years or so. In the Mediterranean though, only 1.6 per cent of fires is due to lightning, 42.2 per cent to negligence, and 14.6 per cent to deliberate firing, the total area burnt in the 1980s being some 600 000 ha per annum, three times the 1960s figure. The number of fires in Portugal has increased by 50 per cent in 30 years, though the area burnt has decreased slightly to 57 000 ha in 1985. Fire would probably have been used by 'hunter-gatherers' to drive animals, and also to encourage the lush growth on which these animals fed. It has been argued that the progenitors of wheat and barley were favoured by such conditions and this could have been the first stage in their domestication in the Near East. Of native plants, *Pistacia lentiscus* is resistant to drought, fire, grazing, and cutting, and remains as the last woody survivor in degraded *matos*. It resprouts from deep extensive underground parts and from layers, forming a dense compact canopy with high soil- and water-conserving features. Its leaves are resinous and soon become distasteful to animals. *Quercus coccifera* resprouts more readily after spring or winter fires than after the more intense summer and autumn ones. All the facultative resprouters are low, unpalatable, or are aromatic shrubs such as species of *Cistus*, *Thymus*, and legumes. Geophytes like *Asphodelus* spp. are completely unpalatable to animals including insects and resist the most intense grazing and fire regimes to dominate even the most degraded of landscapes.

Grazing animals under traditional systems did not eat out the habitat because there was not enough food in winter. Hence the carrying capacity of the habitat was reduced, allowing the spring flush to produce flowers and seeds. Winter-feeding produces bigger animal populations with consequent overgrazing of the spring flush resulting in a failure to regenerate.

Human history

The waves of different peoples entering the Algarve have modified the vegetation, either by the management, inadvertent or otherwise, of the native plants, or by the purposeful introduction of new plants. The wheat, barley, sheep, goat, cattle, and pig agriculture of the east seems to have been established in the west Mediterranean coastlands during the fifth millenium BC, though the only recorded agricultural settlement in Portugal for the period to the third millenium is in the Lisbon area. Intensification of this rudimentary dry-farming agriculture was delayed until the introduction of olive, grapes, and figs from the increasingly sophisticated wheat, oil, and wine economy of the east.

In the middle and late Bronze Ages, olive cultivation was well established in the eastern Mediterranean. The cultivated olives were possibly selected in the mountainous areas of the eastern Mediterranean and, in being moved westwards, there was a second source of diversity in the Aegean and a third even further west in the southern Italy–Tunisia region. By 600 BC the Romans knew of these olives and some of the Tunisian clones familiar to them still survive. Similarly, some fig cultivars are very old, for example, 'Sari Lop' ('Lob Injir' and renamed 'Calimyrna' in the USA) has been grown in Turkey for 2000 years, whilst the Italian 'Dottato' was praised by

Pliny. Portuguese olives seem to have originated from the third centre, some of the trees in the *Sotavento* perhaps dating from the original introductions. Olives and grapes were certainly grown in eastern Spain by 530 BC, and it has been suggested that the Phoenicians and Greeks introduced the orchard system.

Not only the vegetation has been modified: it has been cogently argued that the introduction to the western Mediterranean of cultivated plants for food (and ornament) has changed the migrating and wintering behaviour of fruit-dependent bird species, allowing them to winter in more northerly or different regions than they did formerly. Grapes are important food for thrushes, for example, and the almost exclusively frugivorous blackcap relies more and more on introduced plants as the season moves from summer to early spring. When olive crops are poor in southern Spain and harvesting is therefore unprofitable, as it was in 1981–2 for example, there is in consequence a higher food resource in the orchards than in the surrounding *matos*, leading to a high blackcap density in the former.

Olive stones and vine pips have been recorded at four archaeological sites in central Portugal:

(1) *Gruta de Redondas*, Algar do João Ramos (north of Lisbon), where Neolithic (4500 BC) remains were described by Pinto da Silva;

(2) *Vila Nova de São Pedro*, Cartaxo (between Lisbon and Santarém), where Chalcolithic (2400–2200 BC) remains were described by Pinto da Silva and by Hobf — seeds recovered included remains of wheat and six-rowed barley, broad beans, peas, and linseed;

(3) *Zambujal*, Torres Vedras (west of Santarém), where Chalcolithic remains found included wheat and barley, broad beans and olives;

(4) *Castrum Baiões*, Beira Alta, Viseu, from this late Bronze Age site, seeds of wheat, barley, peas, and broad beans have been described.

There is similar evidence from Spain. A cemetery of the copper-using communities in the Algarve is known; these may have been 'colonists' from the third millenium Aegean but, by 2000 BC, these sites were re-occupied by indigenous people. Fire may have been used in warfare as well as agriculture. Hunting certainly removed much of the fauna, even the carnivores which were probably important dispersal agents for the native plants, notably in southern Spain, for example, *Juniperus phoenicea*.

Agriculture in Iberia was intensified in response to Phoenician, Greek, and later, Roman, demands. The Phoenicians set up trading posts in the twelfth century BC to be followed by peoples from the north of the Pyrenees in the fifth century BC and, at the same time, the Carthaginians arrived from the south. The Greeks visited the coast about 400 BC. Little is known of the Carthaginian occupation but it ended with the conclusion of the Second Punic War in 201 BC. Roman Lusitania was administered from Mérida (*Emerita Augusta*), now in Spain. Algarve and Alentejo were advanced culturally when the Romans arrived and, although the colonists established their large estates and great villas (*vilas*) near Beja (*Pax Iulia*) and Évora (*Ebora*) in Alentejo for example, they also built large villas along the Algarve coast. There are important remains at *Cerro de Vila*, Vilamoura, whilst one of the largest in Portugal is at Milreu and there is another at Quinta de Marim. At Vila do Bispo there are vestiges of Roman iron pyrites mines.

Figure 233 The villa with chapel at Milreu.

Under the Romans, the Algarve was controlled by the *Conventus Pacencis*, an administrative body with judicial powers that extended as far north as the river Tagus (*Tejo*). Balsa (now Luz de Tavira) was responsible for the east of the Algarve, Ossónoba (near modern Faro), or possibly Lagos or Silves, administered the rest. The road system of the Roman empire was of vital strategic and political importance, but that for Portugal as a whole is still incompletely known, partly because 'modern' road surfaces have been laid on Roman, stone, foundations. The Antonine Itinerary (*Itinerário do Imperador Antonino*), first prepared in the third century AD, though subsequently corrected, is a guide marking the roads of the Empire, the main centres of population and the approximate distances between them, as well as the places to stop. Many modern Portuguese towns can be identified and it shows that the main route in the Algarve was from Lacobriga (now Lagos) to Baesuris (now Castro Marim), passing through Ossónoba and Balsa. From Lagos, the road continued to Sagres following the present alignment, and from Lagos northwards it cut through the mountains into Alentejo. The road north presumably went near Monchique, since so many Roman mosaics and other remains have been found in that area, notably at Caldas de Monchique.

During the first millenium BC, Roman gardens at Pompeii were planted with olives, lemons, figs, and citrons, and these orchards were probably intercropped with broad beans and peas. There is evidence that whilst the Romans were in Portugal, they introduced dams leading to reservoirs up to 37.5 km^2 in size, although most of these were less than 3 km^2. They dug artesian wells (*noras*), and hence these wells do not all date from the later Moslem occupation, as is often imagined. The economy was dominated by agriculture, but fishing was important, with the main port at Ossónoba. This city developed as a result of improved relations with North Africa in the third century AD. There were many fish-salting tanks, of which vestiges can still be seen near Portimão, and an *amphora* of *garum* from Castro Marim has been found at Hadrian's Wall in Britain.

Following the Roman decline in the third and fourth centuries AD, a Germanic people, the Suebi, moved south, reaching the Algarve and beyond it, to North Africa, to be followed by the Visigoths. These warlike people were Aryan Christians and they settled in the Algarve from AD 418. By then the Roman system had long been in decline and this was a period of famines, but the Visigoths developed a comprehensive legal code dealing with agricultural matters and, in their time, more forests were cleared for farming.

In AD 711 a force of Moslems crossed the Straits of Gibraltar and, in the following decade, most of Visigothic 'Spain' fell to them. They defeated the Christian King, Rodrigo, making *Gharb* their most westerly province; it comprised most of the southern part of Iberia, and indeed part of North Africa. Even Coimbra and Braga are considered by some historians to have been in the *Gharb*. After AD 1031, with the disintegration of the Cordoba Caliphate, two principalities emerged in the Algarve. These were at Silves (*Xelb, Chelb*) and at Santa Maria de Hárun (formerly *Ossonóba*) which was governed by Ibn Hárun, from whose name present-day Faro is derived. Christian reconquest began with the foundation of the Kingdom of Portugal in AD 1140. Pushed back, the Moslems enlarged and fortified Faro, Tavira, Lagos,

Figure **234**   Silves Castle.

Loulé, and Portimão, and they made Silves their capital, complete with a court which rivalled the splendour of Baghdad. Although conquered in AD 1189 by Dom Sancho I, the first King of Portugal of Silves and the Algarve, it was recaptured, and not finally taken until AD 1249.

Under Moslem rule, agriculture in Iberia seems merely to have been intensified rather than revolutionized. The Calendar of Cordoba (about AD 961) provides the first record of silk and of many new crops: lemon, rice, apricot, cotton, banana, cauliflower, watermelon, and aubergine among them. Sugar and spinach are first listed by al Razi about 930, taro by 1038. Carob is mentioned about 1050, sorghum and indigo by 1100, and probably the pomelo in 1158. Moslems also seem to have brought with them the bitter orange and are believed to have introduced the carrot, the lettuce (*alface*), the almond, and a more efficient form of olive press. Many other words now in the Portuguese language, for example, *aldeia* (a cluster of houses), *alfarrobeira* (carob), *albricoque* (apricot), *alperce* (peach) are Arabic in origin. In the Algarve, there are many Arabic place-names too, among them *Albufeira*, *Alcantarilha*, *Alferce*, *Aljezur*, *Alportel*, and *Alte*. Other towns of Arabic origin are those with the prefix '*Ode*' (water), including the towns of *Odiáxere*, *Odeceixe*, and *Odeleite*, and others with the prefix '*Ben*', as in *Bensafrim* and *Benaciate*. *Portimão* is possibly derived from *Porimunt*, from the Arabic *Burj Munt* meaning 'tower of the farm', that is to say, a fortified agricultural area.

If there was an agricultural revolution under the Moslems, it was a successful incorporation of highly productive tropical cultigens via India into the generalized Mediterranean economy. Of course, this was an intensification and acceleration of a long process of symbiotic adaptation between plants and humans begun long before 'agriculture' was recognizable as such. Eating of figs goes back to pre-hominid days, olive-eating perhaps to pre-primate days, while even the use of cane reflects the 'bamboo-culture' which started in the tropical homelands of the first hominids. In short, all the components of the present-day 'traditional' agriculture as well as cash crops were in place by the end of the Moslem period. Few significant plants have been introduced to that system since, the only prominent novelty in the landscape being the loquat, *nêspereira*, brought to Europe at the end of the eighteenth century.

In the medieval period in the Algarve, the staple foods were meat, fish (often dried), bread, and wine. Although perhaps a rather monotonous diet, the bread at least was of good quality. It was predominantly wheaten, but was sometimes made from a mixture of grains, or from rye or oats only. Daily consumption of bread per person varied between 150 and 750 g. Often there was not enough grain in the Algarve, as elsewhere in Portugal, so wheat was imported to the country from Germany, England, France, Italy, and Spain. There were shortages in some years, which affected the townspeople especially, as people in the countryside had access to other carbohydrate sources, including chestnuts and acorns. With the discovery of the Americas, maize was introduced. The Portuguese name *milho*, originally used to apply to millet, was transferred to the new crop (*milho maiz*). *Milho* is the origin of the English word *mealie* used in Africa for maize.

Before the introduction of the tropical beverage products — tea, coffee, and cocoa — wine and water were the principal drinks, beer largely replacing these in northern

Europe. Portuguese wine, as well as many other agricultural products, was exported to northern Europe, though, as today, the local people consumed most of what was produced. Records of consumption in monasteries of the period show that 1.5 l of wine was allocated daily to each monk: today, Algarvians settle for 1 l per person per day.

At court, and in the richer households, a number of imported goods was used to improve foods: spices from the East, especially ginger and pepper, were readily available. In the thirteenth century, the Royal Family used sugar from Alexandria but, as this cost fifty times as much as honey, the local product was used by the mass of the people, notably for the making of cakes. The production of honey at that time led to the important cottage industry which still persists in the Algarve. Honeycake, *bolo de mel*, better known in Madeira, remains a popular Algarvian speciality. Figs were also widely used for sweetening. Olive oil was used for cooking and as a dressing, although animal fats, especially *touçinho* (lard) were important, as they are still. A number of dairy products were considered to have medicinal value, and butter was much used. Cream and cheese were made in quantity.

Dried fruits were important in the medieval diet: raisins, almonds, walnuts, carobs, and chestnuts in the main. Olives were preserved, and quince and cidra (*Cucurbita ficifolia*) used in 'jams'. Some fruits were not highly regarded however, and Edward (Dom Duarte) who was King during 1433–8, banned the consumption of cherries and peaches, which were considered unhealthy. Many other fresh fruits were eaten, a number of which had been introduced by the Moslems. These included bananas and guavas (*Psidium guajava*), grown especially around Faro. Cane-sugar was grown near Quarteira, perhaps not simply for sweetening, since the stems when chewed have long been used medicinally, as they are known to reduce some types of gastroenterological infection.

By the sixteenth century, there were written recipes using imported spices, such as eels cooked with ginger and saffron and rabbit with cloves as well as ginger and saffron. Wild saffron (*açafrão bravo* from *Crocus serotinus* ssp. *clusii, erva ruiva*), was gathered from the countryside, especially around Lagos, Torre, Odiáxere, and Cape St. Vincent. Over 4000 ha, yielding about 6 kg per hectare of dried stigmas, are still devoted to commercial production of the true saffron crocus (*açafrão, Crocus sativus*) in Spain, where the spice is known as *azafrán*. It is grown mainly around Alcázar de San Juan, Ciudad Real; Carcelen, Albacete; Madridejos, Toledo; Las Pedroñeras, Cuenca; Monreal del Campo, Terual and Cavaca de la Cruz, Murcia. The dye is extracted from the stigmas, a million flowers yielding 10 kg of the spice. It should not be confused with the saffron thistle or safflower (*Carthamus tinctorius*, Compositae, known in Portugal as *açafroa, açaflor, çaflor* and in Spain as *cártamo, azafrán morisco* or *azafrán bastardo*), which was cultivated commercially in the Algarve around Olhão and Lagos into the nineteenth century, usually by sowing between maize plants (see above) in spring. The seed, with yields as high as 1200–1500 kg per hectare, was used for feeding chickens and pigeons, and the saffron-yellow dye was extracted by infusion of the sun-dried flowers in water. The flowers of safflower, as well as those of marigold (*Calendula officinalis*, Compositae), and of arnica (*Arnica montana*, Compositae), an alpine plant, are adulterants of the much more valuable *Crocus sativus*.

Figure 235   Castro Marim in the sixteenth century. (From *Livro das Fortalezas século* XVI.)

Figure 236   Castro Marim today.

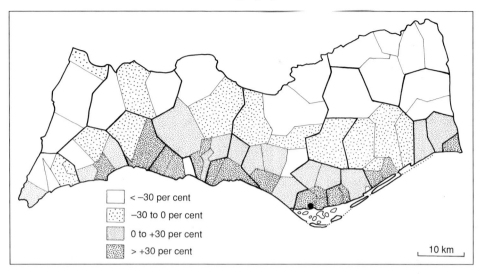

Figure 237   Population growth in the Algarve. *Crescimento demográfico.* (Redrawn from Comissão de Coordenação da Região do Algarve 1985, Anexo B 2.2.)

Important vegetables at this time included the giant cabbage *couve tronchuda*, still the major ingredient of *caldo verde* soups, as well as broccoli, cauliflower, pumpkins, turnips, carrots, radishes, peppers, asparagus, mushrooms, all much as today. Lettuce — although clearly something quite different from the forms now grown — was sometimes served as a speciality with its stalks marinated in sugar (*talos de alface*). To improve the flavour of food throughout the country, vinegar, lemon-juice, and onion-based sauces, as well as olive oil, were widely used.

Perfumes were prepared from orange blossoms and from the leaves and branches of *Cistus ladanifer*. Aromatic oils were extracted from plants such as rosemary, salvia, wormwood (for absinthe), and thyme. Dyes were extracted, especially cochineal, as was the scarlet dyestuff from the abundant kermes oak *Quercus coccifera*. Tobacco had been introduced to the Algarve by the end of the eighteenth century, and was grown at Joinal near Faro. The crop grew easily, and its price was competitive with that from American sources. Laws passed in 1835 extended the production and export of tobacco to Madeira and the Azores.

Demography

Since 1535 the administrative regions have been based on the coastal towns of Lagos and Tavira, followed in 1577 by a bishopric at Faro. The two internationally important ports of Lagos and Faro were largely dependent on the trade in tunnyfish, and Faro and Villa Nova de Portimão (now Portimão) were the two important ports for export of dried fruits (figs, almonds etc.), though Portimão was considered insignificant at the end of the eighteenth century.

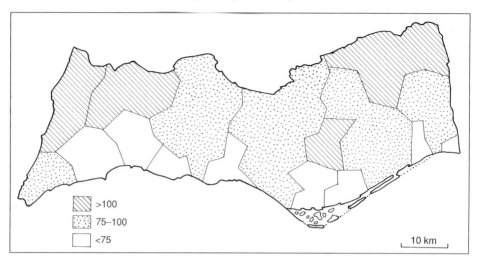

Figure 238   Indices of average age of population. *Indices de Envelhecimento.*
(Redrawn from Comissão de Coordenação da Região do Algarve 1985, Anexo B 2.3)

In 1527–32, Tavira and Lagos, followed by Faro, Loulé, Silves, and Albufeira, had
the largest populations in the Algarve. The Census of 1535 showed 1567 persons in
Tavira, 1310 in Lagos, and 873 in Faro, 536 in Loulé, and 271 in Silves. An estimate
in 1573 indicated that these numbers had increased to 2000 (Lagos), 1800 (Faro), 1500
(Tavira), 800 (Loulé), and 300 (Silves). Thus typical population increases from 1535
to 1573 were about 60 per cent, significantly higher than the European average of the
time. Despite the effects of the plague of 1678, the 1732 census showed a further
increase to 63 682 people in the Algarve, of whom there were 10 235 in Loulé, 7610 in
Tavira, 5117 in Silves, and 3713 in Lagos. By the 1802 census, the population was
105 412 and, in 1837, it had risen to 127 446, with Loulé at 16 943, Tavira 11 801,
Lagos 11 612, and Silves 9767. In 1864, the population was 179 517, in 1920 270 592
and, by 1940, it reached 319 625. The provincial total in 1950 was 325 230, almost the
same level as in 1981 (323 534), although in 1970 it had dipped to 268 957, reflecting
waves of emigration, especially of young workers in the 1950s and 1960s, as a result
of the poor economic conditions then. It was in this period that traditional agriculture
in the Algarve started to decline significantly.

In the *Serra*, there is a relatively high population density around Monchique, with
a corresponding reduction from the average figure for the western *Serra*. There has
been a marked trend, evident from the sixteenth century, to depopulate the *Serra* in
favour of the *Litoral*. In the 1970s and 1980s this tendency to favour the *Litoral* has
increased owing to the reduction of small-scale farming in the interior regions,
combined with the drift of people, both from within the province and from more
northerly provinces, to service the expanding tourist industry which is concentrated
in the coastal zone. Over the same period, there has also been a decrease in
population in areas of the coast affected by changes in the fishing industry and the
decline of the canning factories.

Figure 239 Albufeira at the beginning of the nineteenth century. (From Landmann 1818.)

Figure 240 Albufeira today.

Tourism

The English word 'travel' is derived from *travail* (work) and so the earliest ideas of travel were of it as being something worthy. Touring in Europe for pleasure was developed from the 1840s by men like Thomas Cook. The original attractions of the Algarve for the first tourists, namely those engaged in the Grand Tour of the eighteenth century, were minimal, though they might have visited the ruins of the historic forts in the south-west. Travel by road was difficult, and the sites were so far from other interesting places in Portugal that numbers must have been small. Until the nineteenth century, land communication between Lisbon and the south of the country was bad, being almost exclusively by mule. The main access from Lisbon to the Algarve was by steamboat. Along the Guadiana river, there were also steamboats sailing between Mértola and Vila Real de Santa António. The vessels called at Tavira, Olhão, Faro, Albufeira, Portimão, Lagos, and then followed the west coast of the Algarve to Sines. The railway did not reach Faro until 1889, expanding from there both eastwards and westwards in the early twentieth century.

Indeed Portugal was far from being one of the obligatory places to visit, most of the 'sights' being in Italy; even as late as 1937 Portugal only had an annual average of 36 000 tourists, compared with the five million who went to Italy. And the effect of the 1755 earthquake in destroying much of what was worth seeing in Lisbon meant that few foreign travellers came to the Algarve. However, many Portuguese visited the spas, as had the occupying Romans, notably to 'take the waters' at Caldas de

Figure 241   Portimão Harbour in the early nineteenth century. (From Landmann 1818.)

Figure 242   The Portimão harbour entrance from Ferragudo in the early nineteenth century. (From Landmann 1818.)

Figure 243   The same today.

Figure 244   The bridge at Lagos from the west in the early nineteenth century. (From Landmann 1818.)

Monchique. This mountain spa, near the town of Monchique itself, has long been important. A hospital 'adjoining the waters' had certainly been built there by 1692, and by 1811 this facility had become very well known. At this time, between the intensively cultivated areas around the town were large numbers of chestnuts. In Edwardian times, Caldas became fashionable, but, ironically, in its rebuilding, many of the Roman remains were destroyed.

Apart from Monchique, Baedeker's first guide-book for Spain and Portugal (1898), mentioned only Faro and Loulé; access was by rail. In his 1908 edition, he was less reticent. 'Most travellers will scarcely find it worth while to visit SOUTH PORTUGAL. The towns contain nothing of much interest, while the places on the coast resemble large fishing villages.' For those who would venture there, he advised, owing to the state of the hotels in the Algarve, that travellers be armed with 'insect powder'. By then, though, the railway had reached Portimão and 'diligences' (stage-coaches) conveyed the tourist to Lagos, Sagres, which for part of the nineteenth century was completely abandoned, and Cape St. Vincent. To the east, the line had extended to Vila Real de Santo António, while Castro Marim could be reached from there by boat in half an hour.

Besides taking the waters as at Caldas de Monchique, which in the modern period had been known for their curative properties since at least 1493 when Caldas had been visited by the King, Dom João II, visitors also went to the coastal regions in northern Europe 'for the air'. They carefully shielded themselves from the sun and, as had become fashionable in the reign of the British King George III, went in for 'sea-bathing', and following his example when he visited Cheltenham in 1788, went to

Figure 245   Lagos from O Pinhão in the early nineteenth century. (From Landmann 1818.)

Figure 246   Lagos today.

Figure 247   Silves from the River Arade in the early nineteenth century. (From Landmann 1818.)

Figure 248   Silves today (with a lens it is possible to see the small building in the centre of the photograph just as it was in the early nineteenth century).

Figure 249 Silves from the east early in the nineteenth century. (From Landmann 1818.)

Figure 250 The same today.

Figure 251   Aljezur from the south early in the nineteenth century. (From Landmann 1818.)

Figure 252   The same today.

Figure 253   O Pomar Velho above Monchique early in the nineteenth century. (From Landmann 1818.)

Figure 254   The same today.

Figure 255   Caldas de Monchique with Foia behind early in the nineteenth century. (From Landmann 1818.)

spas. These activities were thought to be good for health. The fad for a tan, which originated in the United States, dates only from the 1920s. In northern Europe, the beaches became crowded with sunbathers and, increasingly, after the Second World War, those seeking exclusiveness sought out the beaches in Italy and the south of France where there were fewer people and more reliable weather than in Britain and the rest of northern Europe. Then, as mass tourism developed, countries such as Spain, Greece, and Portugal became popular, and these are now, in turn, reaching a limit on the expansion of 'basic tourism', so that the fashion of up-market travellers includes going to Caribbean and Pacific resorts.

It was with improved air communications that this mass tourism with all that follows became established so quickly, since the alternative of driving from northern Europe over local roads was a hazardous and lengthy undertaking. The airport at Faro, the only one in the Algarve open to international flights, is on a site of 144 ha, 4 km outside the city, and was not operational until July 1965. By the time a new terminal was opened on 8 August 1989, it was recording 2.5 million passenger movements a year and it is expecting more than 4 million per year by 2000; moreover, the handling capacity could be doubled.

Conservation

The increasing pace of development, particularly for tourism, and the strong probability that European Community pressures will favour the intensification of specific types of agriculture, have caused concern at the local, national, and international level about the effects on conservation and the environment in general. As with many conservation movements, scientists have concentrated on those species which are conspicuous, for example, birds, where they find a large and sympathetic following amongst lay people, or they look at those species unique to an area, taking a 'scientifically respectable' line. Looked at from a provincial point of view, the latter approach can be misleading, since many of the typical Algarvian plants for example can also be seen in Alentejo, because the provincial boundaries do not match ecological ones. Hence, although the level of endemism (plants found only in this province) in the Algarve is low, conservation must be seen in terms of an entire ecosystem in an international context. This is something which the European Community, in conjunction with national and international scientific advisers, is in theory well placed to do. Added to this is the obvious fact that any particular landscape is unique in position and content; once it is destroyed, it is irreplaceable. Much of the Algarve coast with its spectacular cliff formations and the flora and fauna associated with them comes into this category. So do the peaks in the Monchique range and in the Serra da Caldeirão, and the rugged landscapes of the *Barrocal* near Loulé, the island system of the Faro area, the course of the Guadiana southwards from Alcoutim, the extensive dune area at Bordeira near Carrapateira, and the major floristic zone around Sagres–Cape St. Vincent.

As early as 1815, concern was expressed in Portugal over the conservation of forests and, in the Algarve, a plan for the protection of the Sagres–Cape St. Vincent–South-west Alentejo region was made in 1961, but it was only in 1973 that a preliminary list of sites to be protected was drawn up. Seventy-five were identified, eight of them in the Algarve. By 1975, a Secretary of State for the Environment and Natural Resources had been appointed.

Continental Portugal, assisted by the EC, now has the following reserve areas, with differing degrees of establishment and protection, from National Park down to the minor status of Protected Area. There is only one National Park (*Parque Nacional*), which is the montane Peneda-Gerês (600 km^2 at 1545 m). There are six Nature Parks (*Parques Naturais*): Montesinho, *Serra da Estrela* (522 km^2 at 1991 m), *Serras de Aire e Candeeiros*, coastal *matos* on limestone, Arrábida (108 km^2), and Alvão (70 km^2). There are nine Nature Reserves, *Dunas de São Jacinto*, *Serra da Malcata*, *Paúl do Boquilobo*, *Estuário do Tejo*, *Estuário do Sado*, *da Berlenga*, *Paúl de Arzila*, and two in the Algarve, which are the coastal *Ria Formosa* (about 14 500 ha) and the *Sapal de Castro Marim e Vila Real de Santo António* (about 2089 ha). Also planned for the province are Nature Reserves at *Ponta de Sagres* and *Serra de Monchique*. Finally, there are six Protected Areas (*Áreas de Paisagens Protegidas*), at Sintra-Cascais, *Serra do Açor*, *Arriba fóssil da Costa da Caparica*, *Açudes do Monte da Barca e Agoloda*, *Litoral de Esposende*, and lastly the only one in the Algarve, which extends from Sagres up to Sines in Alentejo, that of *Sudoeste Alentejano e Costa Vicentina*, about 70 000 ha of which 25 850 ha are in the Algarve.

The eight areas in the Algarve listed in 1973 were the Lower Guadiana and several other coastal sites, many of which are now within the Ria Formosa Reserve, and the salt-marshes at Pêra for their faunas, Serra de Monchique and part of Serra

Figure 256   Rocha da Pena.

da Caldeirão as well as the coastline from Porto Corvo to Cape St. Vincent for their flora and landscape as well as fauna, and part of the *Barrocal* (Vale do Murtinhal, Serra da Arade) for its flora and fauna. Since 1973, the histories of these areas have varied, for, of the country's Reserves, those in the Algarve are all coastal: the *Sapal de Castro Marim e Vila Real de Santa António*, and the *Área Protegida de Costa Vicentina e Sudoeste Alentejo* ('*Costa Sudoeste*') established in 1976 and in 1988 respectively. However, the marshes of Vale de Parra near Armação de Pêra were bulldozed in 1987 to make a golf-course. Parts of the *Barrocal*, for example Rocha da Pena and the area around Monchique, are still being actively proposed for protection. Despite the status of Protected Area for the *Costa Sudoeste*, many developments connected with the tourist industry are still proceeding with scant respect for the environment.

The Ria Formosa Natural Park includes some 3400 ha of 'pré-parque' and a series of interlinked islands and lagoons. The Park exceeds 55 km in extent from Ancão, west of Faro Airport, to Manta Rota, west of Vila Real de Santa António in the east. It is very important for its flora and bird-life, at least 168 bird species having been recorded. About 7500 people live within its boundaries; the majority of them are involved in occupations associated with its river-system, that is to say, the fishing, shellfish, and salt industries. Quinta de Marim forms part of the reserve and has demonstration traditional houses etc. of the *Sotavento*. Notable reptiles at Ria Formosa

Figure 257   Chameleon. (From Crespo and Oliveira 1989.)

include *Chamaeleo chamaeleon* introduced in about 1918 probably by the Algarve workers from the south of Spain and Morocco. There are about 55 species of fish, 62 of bivalves, 35 gastropods, and 22 decapodan Crustacea. These had a sales value in 1989 of US $44 million for fish, US $40 million for shellfish, with 46 000 t of salt being valued at about US $0.8 million. 80 per cent of Portugal's exported bivalves are from the Ria Formosa region.

Apart from the three designated reserve areas, the Algarve now has a total of 39 'biotopic' areas out of over 400 in mainland Portugal (of which some 270 are important in a European context), defined by the Corine satellite programme in each of which specific considerations make them of special scientific interest. These areas total 119 237 ha, which is 24 per cent of the Algarve, 14.8 per cent being in the *Sotavento* and the rest in the *Barlavento*. In mainland Portugal as a whole, 5.6 per cent of the land surface has some degree of protection, thus occupying a fairly low position in an EC context between the extremes of West Germany with 28.2 per cent of its land-surface protected, and Eire with only 0.4 per cent.

There has also been considerable concern over the fate of the unprotected freshwater wetlands of the Algarve, mainly because of their ornithological interest as stopping-places for migrating birds and sanctuaries for resident ones. In western Algarve, 85 per cent of salt-marshes and associated creeks have been reclaimed in the last century, mostly in the period 1940–60, and almost all freshwater wetlands have disappeared. There were originally about 8000 ha of wetlands of which 1900 ha were saline in marshes and estuaries, and 4500 ha were riverine or flood-plain soils but, by 1987–8, this total had been reduced to 2500 ha of which estuaries were 730 ha, salt-marshes 270 ha, the rest mainly *salinas*, irrigated ricefields, and wet pastures. There is now an admirable initiative to give nature reserve status to the A Rocha peninsula with its wetlands in the Alvôr estuary.

The present position of the Algarve and its future

To recapitulate, the Algarve has a total area of *c.* 5000 km², divided into three main areas.

1. *Litoral.* This zone is now largely taken over by tourism with most of its fertile soils, which are excellent for agriculture, given over to that or being held for development. The region contains the greater part of the Algarve's population of about 324 000. On the southern coast, there is an important fishing industry.

2. *Barrocal.* This is about 130 000 ha of mostly limy soils and is the transition zone between the coastal *Litoral* and the montane *Serra*, and where most of the vestiges of the traditional agriculture of the Algarve survive.

3. *Serra.* This is about 320 000 ha, the northern part of the province bordering southern Alentejo; it is largely covered with forest, and has a low population density.

The climate is basically 'Mediterranean' but with some subtropical features (see Chapter 1). Long dry summers (about 3000 hours of sun per annum) alternate with wet winters (averaging 600 mm rainfall a year). Industry and manufacturing in the Algarve on anything other than a very small local scale is almost non-existent, save for the production of rock-salt, sea-salt (see Chapter 2), limestone for cement, and the quarrying of stones and marble-type materials.

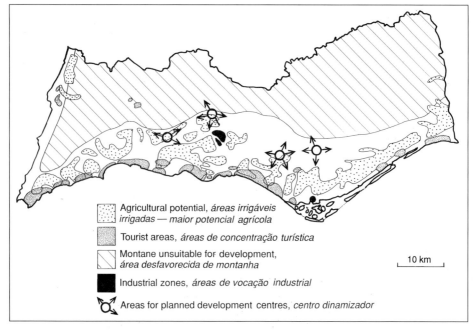

Figure 258 Planned potential of the Algarve. *Principais caracteristicas do uso e occupação do solo.* (Redrawn from Comissão de Coordenação da Região do Algarve 1985, Anexo B l.l.)

Traditional agriculture

As to present-day agriculture in the Algarve, the major change (see Chapters 4 and 5) since the early 1900s, much of it since the early 1960s, has been the virtual eradication of the fig industry, annual production in 1953–62 dropping from about 16 000 t — as dried figs — to 1000 t today, with land taken out of farming completely or planted with citrus. As a result, annual citrus production in the Algarve has now risen to over 150 000 t from 16 000 t in 1953–63. Of the other traditional crops, annual almond production slumped from 10 000 t in 1953–62 to 1500 t.

Carob has drifted down from 40 000 t in 1953–62 to about 30 000 t but remains important owing to the greatly enhanced value for its new-found industrial uses. Olive oil production in the Algarve has declined by about 75 per cent from its level of the 1950s, with the oil now of lower average quality (higher acidity). The 1990 estimated production from about 600 000 trees, averaging 20 kg of fruit per tree, is about 12 000 t, which is some 12 million l, compared with some 40 million l, typical of the early 1900s to 1950s. Production of almonds, figs, and olives continues to diminish gradually with few new plantings and poor maintenance of what remains.

The dried fruits — figs, almonds and carob and their derivatives — have been the economic basis, along with small animals — mostly goats, pigs, some sheep, a few cattle, and with mules or donkeys for power — of the Algarve for many centuries. In much of the *Serra*, traditional agriculture has been accompanied by handcrafts such as basket-making. Small pockets of farmers will probably remain for many years to come, but the major crops are likely to change over the next 20 years.

Figs are likely to continue their decline, although there is no reason why the Algarve should not once more become a major producer of quality fruits in large volumes.

Carobs are likely to stay around their present level. Production, especially in the *Barrocal*, could easily be increased, and it will be worthwhile to do so because of the

Figure 259 Shepherd.

Figure 260   Horse at a fair in Portimão.

Figure 261   Windmills below Rocha da Pena.

Figure 262   Abandoned irrigation system near Abicada.

lack of world competition and because of the high-added value products which can now be derived from the fruits.

Olives are likely to decline in both quantity and quality, with little possibility of reversal, since in view of overall Community surpluses the EC is now paying subsidies for grubbing out olive trees, whilst not replanting with more productive ones of better cultivars and introducing modern cultural methods.

Almonds are likely to decline gradually further, as the economic disparity between the Algarve and other areas in the USA (*EUA*) and Spain, where large-scale plantings have been mechanized to a high degree, is now probably too great to overcome.

Grapes have been reduced to negligible numbers since very small-scale production has become increasingly difficult owing to the labour involved. The average quality of the vines is poor, and they require much spraying to control mildews and pests. Moreover, larger producers are encouraged by the EC to produce higher-quality grapes in demarcated zones. Thus there will probably be a decline in wine grown for everyday purposes, though this could easily be more than counterbalanced by substantial increases in the production of wines of higher value, not only demarcated qualities but wines peculiar to the Algarve such as *Algar Seco* (Lagoa), *Moscatel* (Lagos), and the *Dom Paio* of Tavira. Furthermore, there is a major potential in greatly increasing cultivation of first-quality table grapes (*uvas de mesa*). Production is easy, quality can be controlled, and the marketing system is already largely in place; all could easily be expanded.

Eucalyptus

Ecological change throughout the Mediterranean has led to the resurgence of forest as farming is neglected. Whereas, in the late eighteenth century, the Portimão region was noted for its cornfields with olives and figs in them, the area, like so many of the countries north of the Mediterranean, is now characterized by shrubby formations (*matos*), whilst in the *Serra* to the north, eucalyptus is prominent.

Notwithstanding mounting environmental objections, it is likely that the pace of eucalyptus plantings will increase in the *Serra* and that these trees, which can produce cellulose so rapidly, will be the only significant crop in those areas. We have pointed to the dangers inherent in the monoculture of eucalyptus (see Chapter 6), with evidence already accumulating that the trees are being seriously damaged by *Phoracantha*. It is essential to maintain a balance of tree species and in similar places we expect to see some plantings of pine and cork oak and other oaks, but few walnuts. Nevertheless, forestry in the Algarve is advantageous in that the EC has a wood deficit, which will probably not be covered in the near future. It is not expected that there will be any significant increase in plantings of the strawberry tree, although this shrub is very suitable for much of the higher altitude areas of the Algarve. The harvesting and processing of the crop by traditional methods are labour-intensive, and even the present stands are poorly maintained.

Intensive crops

Although much of the land between the *Litoral* of the tourists and the degraded shales and syenite of the *Serra* is irrigated, either from reservoirs in the mountains, or from local boreholes, it is essential to introduce a scientifically based integrated control of all water resources (see Chapter 1). Fresh fruit crops (*frutos frescas*), especially citrus, occupy quite a large total area, but the orchards are divided among many producers. Citrus grows very well and, indeed, has been a valuable crop in the Algarve for centuries (see Chapter 5). However, the present production price is high by world standards; considerable investment is needed to produce quality fruit at competitive prices. This requires major structural changes, including increased and replacement plantings of virus-free citrus on disease-resistant rootstocks on well-managed plantations, orange and grapefruit cultivars for juicing, and replacement plantings of citrus which mature 'off-season'. Appropriate diversification could include apricots, which can be grown very easily where there is now low-quality citrus in the Algarve. Loquats could be planted on a much larger scale, and with great commercial benefit, especially since irrigation is unnecessary, though more appropriate cultivars, known in Spain, need to be introduced.

With the equable climate in the Algarve and sufficient water for irrigation in some 50 000 ha, there is undoubtedly much scope for many more tropical and subtropical foodplants. Major increases in the production of avocado pears (see Chapter 5) are feasible and could generate desirable returns. Already grown on a smaller scale, but scarcely marketed, are the fruit trees of South American origin, soursop (*anoneira*, *Annona muricata*, Annonaceae) and the cherimoya ('custard apple', *A. cherimola*). Custard apples can be grown only on acid soils up to 300 m, so they are rarely seen either in the high *Serra* or in the *Barrocal*. Other food-plants which could be grown on a plantation scale in the province include passion fruit (*maracujá, Passiflora edulis*), guava (*Psidium guajava*, Myrtaceae) — particularly useful since it ripens in November and December — litchi (*lichi, Litchi chinensis*, Sapindaceae, and grown on Spain's southern coast near Malaga since 1976), longan (*Dimocarpus longan*), kiwi-fruit (*Actinidia deliciosa*), pineapple (*ananazeiro, Ananas comosus*, Bromeliaceae), papaya (pawpaw, *papaieira, Carica papaya*), pistachio nuts (*pistácio, Pistacia vera*), and the mango (*manga, Mangifera indica*, Anacardiaceae), which has been grown for over 4 000 years in India and was introduced to Brazil by the Portuguese. By weight, the value of all these fruits is about four times that of citrus; with an established hard-currency north European market for exotic fruits, there are clearly economic advantages. Although *estufas*, now covering only a little over 500 ha, mostly in the *Litoral*, will be needed for these crops, no heating is necessary in the Algarvian climate, so that production costs are much lower than in northern Europe. Additional plantings of disease-resistant cultivars of chestnuts (see Chapter 6) could generate worthwhile returns.

Other commercial possibilities in such areas include increased production with modern techniques of cutflowers and exotic pot plants, as well as fresh vegetables out-of-season, for example, green beans and tomatoes, broad beans and peas, potatoes, and soft fruits, especially strawberries. Peanuts (*amendoim*) are another readily grown crop with good market potential, though perhaps not on a world-scale owing to American competition. Of vegetables very easily grown under Algarvian conditions without watering, globe artichokes (*Cynara scolymus*, Compositae) and 'Jerusalem' artichokes (*Helianthus tuberosus*, Compositae) are of considerable interest.

Nevertheless, significant changes in land use have come about in the Algarve only since the early 1960s, coinciding with the onset of tourism on an ever-increasing scale. In those three decades, the major 'loser' has been small-scale integrated farming, not only because it seemed that short-term capital gains provided advantages not seen with agriculture, but also because the bulk of the labour required for tourism and associated activities has come from the younger people of the Algarve, who not unreasonably prefer the higher immediate returns from tourism to the harder manual work and uncertainties of the small farm. The onset of the democratic process in Portugal (from the Revolution of 25 April 1974) taken with the weakening of traditional values, not least religious belief, has given this process even more impetus. Infrastructure projects, especially the building of new roads and the widespread introduction of telephones, piped water, drains, centralized waste disposal, and so on, have to some degree removed the necessity for small groups of the Algarve's people to remain totally integrated within the scale of their own villages. Furthermore, the political practicalities of the EC of which Portugal, with Spain, has been a full Member since 1 January 1986, have brought it about that there is probably no long-term place for small farmers unless they specialize, nor for the crops which they grew so well largely for their own use, for barter or for small local markets.

And so the production of crops such as figs and almonds has fallen dramatically, and remains in steep decline. Fishing (concentrated on the ports of Olhão and Portimão) from small boats, another ancient practice, is stagnating, and problems are looming in the shellfish sector too, owing to uncontrolled pollution of feeding

Figure 263   Cliffs and high-rise hotels near Praia da Rocha.

Figure 264   Unspoilt cove with towers of Torralta beyond.

Figure 265   A Rocha, development or reserve?

Figure 266   The Alvôr Estuary being dredged and developed.

Figure 267   The sand-dunes at Carrapateira with planning permission for a caravan park.

grounds. None the less, as with agriculture, provided that there is adequate investment and sound technical advice and training, it is clear that the fishing industry,
especially that based on high-value shellfish and on aquaculture, is capable of
significant development within the confines of Community quotas. The associated
packing industry, notably of sardines, has declined, but there is scope for it to be
revitalized for shellfish.

Tourism and conservation

The influx of many foreign residents and the rise of the tourist industry, and now the
investment of EC monies into the region, mostly for 'infrastructure' connected with
tourism, and totalling some US \$400 million, has had the greatest effect on the
landscape of the Algarve since the great earthquake of 1755. In Portugal as a whole,
tourism now contributes 9 per cent of Gross Domestic Product (GDP), of which
about half comes from the province, and it represents 27 per cent of foreign
investment. Although the main tourist zones are pockets on the southern coast of the
Algarve, it must be emphasized that the effects of tourism are spread throughout the
province. A major consequence is that the socio-economic system which prevailed
for centuries until the 1960s has now gone for ever. To use but one measure, some
60 per cent of the Gross Internal Product of the province is now generated from
tourism. Following the growth of tourism has come a very large programme of road-
building, culminating in the planned Via do Infante motorway — cutting laterally
across the Algarve — being built (1992) on a line somewhat north of, but otherwise
approximately parallel to, the existing EN125 highway, which will provide a new
major link to Spain.

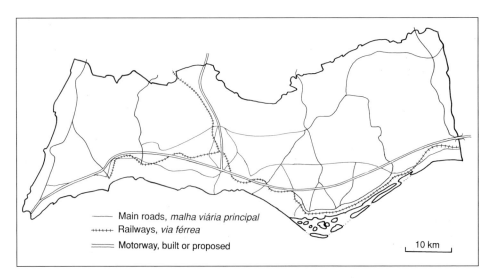

Figure 268   Major road and rail network in existence or planned in the Algarve.
Malha viária principal. (Modified from Comissão de Coordenação da Região do
Algarve 1985, Anexo B 1.2.)

PROTAL

The *PROTAL* 'structure plan' (*Plano Regional de Ordenamento do Território*) for the Algarve, published 26 January 1989, is a comprehensive planning document for the province, including the southern coastline, except around Lagos which escaped this control as the earlier plan of that *concelho* had been approved. *PROTAL* has been given 'teeth' by the *REN* (*Reserva Ecológica Naçional*) Law of 19 March 1990, which appears to give wide executive planning powers to the regional co-ordinating body, *Comissão de Coordenação da Região do Algarve* (CCRA), which until now had purely advisory functions. None the less, it is difficult to say now how effective this overall plan will be, since the *câmaras* continue to exercise very considerable autonomy in planning as well as in other local matters, and there is a large backlog of existing consents. Orderly planning leads, *inter alia*, to the possibility of a more stable and lucrative tourist industry.

It is likely that tourism will continue to dominate the scene for some years, with all else effectively being subsidiary to it. This will be so particularly in the *Litoral* coastal zones in spite of the protection apparently afforded against further development. Because of the labour demands of the tourist industry, where wages are much higher than those in traditional agriculture, it is inevitable that the old socio-economic system of small farmers throughout the Algarve, the majority working less that 5 ha of land, will continue to be broken up. The serious effects of this will not be seen for a generation, since much of the agriculture is being kept going by the older people, while the young move into tourism, the construction industry, and so on. National figures show that of the active population working in agriculture, only 8.6 per cent are less than 35 years old, whereas 25 per cent are over 65 years old. Ominously, however, there is evidence of a decline in arrivals by air from some countries and of a levelling-off in housing and general construction, indicating that the basic tourism of the 1970s and 1980s has peaked already.

Pressures for conservation continue to rise and may now also be coming from an unexpected quarter. Tourists, particularly of the 'quality' type wanted by the authorities, today expect more than sun, sand, and cosmopolitan food. The Algarve has a great potential for high-quality tourism in its people, its monuments ranging from Neolithic to medieval, its varied landscape and protected areas, its beautiful flora and its fauna as well as its geographical position and climate. These factors could be exploited by a return to more 'classical tourism'. To do this it is essential, as *PROTAL* envisages, to permit developments only in an integrated economic context of the whole province and to respect the designated conservation areas (see above).

These sentiments apply to the wild organisms and scenery of the region. But this, as we have seen, is the result of thousands of years of management by human beings; much of the modern landscape can only be maintained if the traditional land-use methods are continued. It seems highly unlikely that such farming and the social system of which it forms a vital part — with the integrated uses of native and exotic plants and animals and the quality of village life — can co-exist on a significant scale with EC policies and the demands of tourism. The changing social pattern is now effectively irreversible. Moreover, the small size of the farms and the independence of their proprietors means that modern machinery, even if made available, could not, in any case, easily be used to raise productivity. Nevertheless, it is essential that deep thought be given to this dilemma and help provided, not only to provide a buffer against any major decline in 'basic tourism', but to respect and build on the invaluable agricultural tradition of the indigenous people of the Algarve.

The future

What happens now that basic tourism is stagnating, not only in the Algarve, but in many other European resorts? Nothing can be done locally about the world economy and trade, but clearly it would be a wise insurance policy for the Algarve to make the best provisions it can. This could include, for example, the return to more classical tourism and a sounder form of traditional agriculture, coupled with the balanced introduction of high-value crops. In addition, the closer economic union with Spain has to be developed, not simply as two fledging members of the European Community, but because Portugal and Spain have their own 'Iberian economic union', which provides a wider market for the capital, goods, and services which the Algarve has to offer. However, because of this link, it is even more important than before that the province develops 'niche markets' both to differentiate it from its much larger neighbour and to learn from the advances, and sometimes mistakes, which Spain has made.

It could be argued that the future of the Algarve is a rosy one provided that scientifically-based programmes are developed and implemented in the near-term, making proper use of EC 'structural funds' whilst there is time to do so before Portugal is fully exposed to competition from its strong fellow Members, and taking

Figure 269 The end. (Photograph Travelpress Europe.)

overall regional, national, and European needs into consideration. These programmes include

- making it worthwhile for more of the younger *Algarvios* to stay in the Algarve and to encourage others to come and settle in the province,
- halting, and then reversing, the steep decline in traditional agriculture, bolstering this with a major restructuring of the citrus industry and the introduction of suitable sub-tropical fruits,
- introducing an integrated programme of expansion of the fishing industry and its marketing, including high-value aquaculture, and
- setting up schemes to develop higher-quality tourism, coupled with the necessary infrastructure, which in turn needs to be integrated into
- a programme for the conservation of the flora and fauna of the Algarve in the context of its rich heritage of historic monuments and the unique culture of the province and its people.

Sources

References by chapter

General

Amaral Franco 1971–84; Amaral Franco and da Rocha Afonso 1982; Anderson 1956; Anderson and Anderson 1991; Angeles Mendiola 1989; Anon. 1973*a*, 1990*o*, 1991; Backeberg 1976; Baumann 1984; Beckett 1988; Bonnet 1850; Bramwell and Bramwell 1974; Brazão Gonçalves 1988; Castroviejo *et al.* 1986–90; Cavaco 1976, 1983; Costa and Franquinho 1986; Costa Primo 1936; Coutinho 1915, 1939; Crespo and Oliveira 1989; Cyrén 1930–1; Daveau 1988; Innes 1987; Jones 1978; Landmann 1818; Lanza and Pizzetti 1979; Le Houérou 1974, 1981; Link 1801; Mabberley 1990; Macedo and Sardinha 1985; McMurtrie 1973, 1986; Malato Beliz 1990*a,b*; Margalef 1985; Marques Moreira 1987; Piggott 1965; Pinto da Silva 1963; Placito 1991; Polunin 1969; Polunin and Everard 1976; Polunin and Smythies 1973; Polunin and Walters 1985; Rego *et al.* 1989; Rocha Afonso and McMurtrie 1991; Schönfelder and Schönfelder 1990; Silva Ferreira Sampaio 1949, 1988; Stevenson 1985; Taylor 1972; Tutin *et al.* 1964–80; Valdés *et al.* 1987; Vedel 1978; White 1983; Wuerpel 1974.

Chapter 1   (Geography and ecology)

Anon. 1959, 1977–9, 1982*a,b*, 1988b; Beato 1982, 1986; Bordalo da Rocha 1976; Bordalo da Rocha *et al.* 1979; Botelho da Costa 1985; Cabral *et al.* 1977; Carvalho Quintela *et al.* 1987; Casimiro Martins *et al.* 1992*a,b*; Chester and James 1989; Cornaert *et al.* 1988; Correia 1982; D'Almeida and De Almeida 1966; Daveau 1977; Daveau *et al.* 1985; Davis *et al.* 1986; Devereux 1982; Dias 1986; Feio 1983; Folques 1987; Gaspar 1981; Instituto Geográfico e Cadastral 1964; Instituto Nacional de Investigacão Científica 1989; Kopp *et al.* 1989; Lautensach 1988; Lopes Leitão 1992; Madeira 1969, 1984; Manupella *et al.* 1984; Martin *et al.* 1984; Mimoso Loureiro 1988; Mimoso Loureiro and Fonseca Nunes 1983; Moreira 1989; Moreira *et al.* 1989; MPAT 1989; Nuncio *et al.* 1988; Parker 1981; Pena *et al.* 1985; Ramos *et al.* 1988; Reis Cunha 1957; Ribeiro *et al.* 1987; Rocha Faria *et al.* 1981; Romariz and Andrade 1989; Sérgio *et al.* 1984; Serviços Geológicos de Portugal 1970; Sutton 1988.

Chapter 2   (Coastline)

Akeroyd and Preston 1990; Alves de Araújo 1987; Anon. 1986, 1989i; Azeredo *et al.* 1969, 1984; Bolton 1989; Correia Peixoto 1991; Costa and Lousã 1986; Crawford 1989; Cruz 1986; Daveau 1897; Duarte *et al.* 1984; Espirito Santo 1989*a,b*, 1990; Faleiro Silva 1989; Fritz *et al.* 1988; Gómez-Campo *et al.* 1984; Gonçalves Vieira 1911; Lebreton and Rivera 1988; Lepierre 1935; Luther and Fiedler 1976; Morais and Carvalho 1992; Moreira 1978; Moreira 1989; MPAT 1989; Preston and Sell 1988; Pullan 1988*a,b*; Quéro *et al.* 1990; Rau 1984; Rosa Santos 1989; Rothmaler 1943; Ruano *et al.* 1987; Secretaria de Estado do Ambiente 1988*a,b,c, d*; Serrano 1988; SNPRCN 1986; Tavares 1961; Tavares and Saccarão 1960; Woodell 1989; Woodell and McDonald 1989; Zedler and Scheid 1988.

Chapter 3 *(Lowlands)*

Borg-Karlson *et al.* 1985, 1987; Crawford 1989; Davies *et al.* 1988; Dias *et al.* 1989; Espirito Santo 1989*c,d*; Goldblatt 1980; Guillerm *et al.* 1990; Henriques 1890; Herrera 1981, 1982, 1984, 1985, 1986, 1987, 1988*a,b,c*; Ingram, unpublished results; Izhaki and Safriel 1989; Jordano 1982, 1989; Juhren 1966; Kullenberg and Bergström 1976; Malato Beliz 1986; Martin and Escarre 1980; Michael 1964; Montalverne 1988; Montgomery and Strid 1976; Montoya Oliver 1990; Naveh and Whittaker 1979; Ornduff 1987; Orwell 1938; Owen 1980; Perez Bueno 1989; Placito 1986, 1987; Proctor and Yeo 1973; Rabaça 1988; Service 1989; Taborda de Morais 1938; Taylor 1990; Thomas 1979; Trabaud 1977, 1983; Trabaud and Chanterac 1985; Trabaud and Lepart 1980; Vuillemin and Bulard 1981; Warburg 1931.

Chapter 4 *(Traditional agriculture)*

Almeida 1942, 1984; Anon. 1850, 1898, 1983, 1987*a*, 1989*h*, 1990*e*; Barranco and Rallo 1984; Bivar Gomes da Costa Weinholtz 1883; Boardman 1976; Bravo de Lima *et al.* 1988; Cabreira 1918; Cabrita Neto 1980; Campos Pereira 1915; Carneiro and Bravo de Lima 1987; Catela Pais Mousinho 1984; Cavaco 1983; Chiej 1984; Clutton-Brock 1987; Collenette 1988; Condit 1925, 1947; Corrêa da Mello Leotte 1901; Curvelo-Garçia *et al.* 1987; D'Almeida de Bivar Weinholz 1891; Ferreira de Mira 1939; Figueiredo e Meneses 1970; Flores Dominguez 1990; Galil and Neeman 1977; Garcia Alonso 1990; Gardé and Gardé 1988; Gimeno 1987; Grasselly and Crossa-Raynaud 1980; Green and Kupicha 1979; Green and Wickens 1989; Guerreiro Costa 1977; Guerrero 1991; Hadlington 1990; Hartmann *et al.* 1980; Heitor and Fernandes 1989; Hills 1980; Ibar Albiñana 1985; Jones 1980; Kjellberg *et al.* 1987; Leach 1982; Leibold 1980; Lopez Palazon 1970; Loussert and Brousse 1978; Luna Lorente 1986; Mabberley 1991; Martins Loução and Brito de Carvalho 1989; Mattos Parreira 1919; Meith *et al.* 1983; Mira Galvão 1939, 1940; Miranda 1909; Modesto 1988; Neto Martins and Branco 1990; Neves 1985; Obenauf *et al.* 1978; Oliveira Marques 1978; Palmeirinha Godinho Rebelo and Couto da Rocha Pita 1992; Parker and Stevens Cox 1970; Pato 1988; Payne 1968; Pemberton and Turner 1989; Pereira de Mattos 1900; Ribeiro 1980; Ripado (n.d.); Rivera Nuñez and Walker 1989; Robinson 1986; Salvador and Ramos 1989; Santos Nascimento 1944; Simmonds 1976; Soares 1965; Sousa Pontes 1982; Spina 1989*b*; Stevenson 1985; Tous Marti and Batlle Caravaca 1990; Trancoso Vaz 1955, 1987*a,b*; Unwin 1991; Veiga de Oliveira *et al.* 1983; Vianna e Silva 1983; Vitoriano 1980; Winer 1980; Zohary 1982; Zohary and Hopf 1988.

Chapter 5 *(Citrus and other cash crops)*

Alba Ordoñez and Llanos Company 1990; Alsaadawi *et al.* 1985; Anon. 1980, 1981, 1984, 1990*g*; Bowman 1956; Buczacki and Harris 1981; Cobianchi *et al.* 1989; Costa and Entrudo Fernandes 1990; Cunha Mota 1919; Duarte Amaral 1982; Entrudo Fernandes and Grilo 1989; Ferguson *et al.* 1990; Ferreira 1990; Fideghelli 1987; Furtado 1990; Gaillard 1987; Galan Sauco 1990; Ibar Albiñana 1986; Iglesia Gonzalez 1988; LaRue *et al.* 1983; Lima Miranda 1974; Luna Lorente 1990; Madeira Lobo 1977; Marota Borrego and Lopez Galarza 1988; Moreira da Silva & F*os* 1982; Neto Martins and Branco 1990; Nolasco 1988; Paglietta 1986; Palhavã 1988*a, b*; Passos de Carvalho 1986; Purseglove 1968–72; Ramos 1985; Reche Marmol 1988; Reuther *et al.* 1967–89; Samways 1983; Simmonds 1976; Simmons 1972; Soares Chaves 1980, 1988; Sousa Veloso *et al.* 1988; Tamaro 1984; Teixera Gomes 1986; Zapata *et al.* 1989.

Chapter 6 *(Serra and freshwater)*

Afonso 1990; Amorim 1989; Anon. 1959, 1964, 1989*a,d*, 1990*c,d,m,p*; Barros *et al.* 1986; Cooke 1961; Cruz Rosa 1978; Damásio 1990; Davidson and Jeppe 1981; Debazac 1983; Duarte *et al.* 1984; Emilia Rosa and Amaral Fortes 1987, 1988*a, b, c, d, e, f*, 1989; Espirito Santo 1989*e*; Estácio da Veiga Silva Pereira 1866; Fabião 1987; Feio 1989; Fonseca Leal de Oliveira 1990; Gascon 1955; Horner 1990; Juniper *et al.* 1989; Kardell *et al.* 1986; Le Houérou 1987; Macedo and Sardinha 1985; Malato Beliz 1982; Meiggs 1982; Monteiro Alves 1988; Montoya Oliver 1988; Parsons 1962; Pereira *et al.* 1987; Pessoa 1988*b*; Poore and Fries 1985; Rebelo and Rito (n.d.); Reimão and Nunes 1989; Seigue 1985; Silva Carvalho 1939; Silveira da

Costa 1985; Soares Barreto 1988; Stanislawski 1962; Stowell 1989; Vasconcelos Lima and Gomes de Castro 1988; Wojterski 1990.

Chapter 7   (Streets, parks, and gardens)

Chittenden 1951; Cunha Mota 1919; Goes and da Silva 1986; Graf 1978; Herklots 1976; Jacobsen 1960; Latymer 1990; Mabberley 1984; Noailles and Lancaster 1977; Wrigley and Fagg 1983.

Chapter 8   (Past and future)

Alarcão 1988; Anderson 1956; Anon. 1954, 1959, 1973*b*, 1987*b*, 1988*a*, 1989*b,c,e,f,g*, 1990*a,b,f,h,i,j,k,l,n*; d'Ataide Oliveira 1908, 1913, 1914, 1986, 1987*a, b, c, d*, 1989*a, b,* n.d.; Baedeker 1898, 1908; Barbero *et al.* 1990; Bugalho Semedo 1988; Butzer *et al.* 1985; Cabral Rolo and Seita Coelho 1988; Carvalho de Vasconcelos 1988; Casadinho Parrinha Duarte Regato 1989; Castro Henriques *et al.* 1990; Colino Sueiras *et al.* 1990; Comissão de Coordenação do Região do Algarve 1985; Condit 1947; Correia 1983; Costa Guedes 1988; Davis *et al.*1986; Debussche and Isenmann 1990; Dias 1986; Dray 1985; Crespo and Oliveira 1989; Eliseu 1982; Faustino Crespo 1968; Ferro 1990; Folving and Mégier 1989; Ford and Placito, 1989, unpublished results; Galego and Daveau 1986; Garcia and Cunha 1990; Garcia Rivas 1990; Gomes Gurreiro 1977; Gómez-Campo and Malato Beliz 1985; Graça Silva 1989; Grandjot 1965; Guerreiro 1988; Guilaine 1979; Herklots 1972; Herrera 1989; Leach 1982; Le Houérou 1987; Lonsdale 1980; Loureiro 1984; Louro 1986; Lousã *et al.* 1986; Macário Correia 1984; Marques Ferreira 1989; Martins 1990; Matos 1980; Matos Fortuna 1984; Mendóça Frazão 1992; Mimoso Henriques Cunha 1989; Morais Arnaud and Campos 1986; Moreira and Oliveira 1987; Naveh 1975; Noonan 1989; Nunes 1985; Oliveira Marques 1985, 1987; Paquete 1988; Pessoa 1988*a*; Pinto 1984; Pinto 1987; Pinto Peixoto 1987; Pons and Quézel 1985; Pullan 1988*c*; Ragazini 1985; Raposo 1986; Reis Moura 1989; Ribeiro *et al.* 1987–89; Rodriguez de los Santos *et al.* 1986; Romero Magalhães 1970, 1988; Rufino 1989; Samson 1980; Santos 1984; Serrão 1986; Silva Lopes 1988; Spina 1989*a*; Strasburger 1906; Taylor 1977; Tello Queiroz 1992; Trabaud 1982, 1987; Veiga Fereira and Leitão (n.d.); Viegas Guerreiro and Romero Magalhães 1983; Vilhena Mesquita 1986; Wojterski 1990; Zucherelli and Zucherelli 1990.

Bibliography

Afonso, G. (1990). Novos produtos de cortiça em perspectiva. *Agricultura 92*, **21**, 36–7.

Akeroyd, J. R. and Preston, C. D. (1990). Notes on some Aizoaceae naturalized in Europe. *Bot. J. Linn. Soc.*, **103**, 197–220.

Alarcão, J. de (1988). *Roman Portugal*. 4 volumes. Aris & Phillips, Warminster.

Alba Ordoñez, A. and Llanos Company, M. (1990). *El cultivo del girasol*. Agroguías Mundi-Prensa, Madrid.

Almeida, F. J. de (1942). *A alternância da produção na oliveira*. Junta Nacional do Azeite, Lisbon.

Almeida, F. J. de (1984). *Sobre a reestruturação da olivicultura em Portugal*. Ministério da Agricultura, Comércio e Pescas, Lisbon.

Alsaadawi, I. S., Arif, M. B., and Alrubeaa, A. J. (1985). Allelopathic effects of *Citrus aurantium* L. II. Isolation, characterization, and biological activities of phytotoxins. *J. Chem. Ecol.*, **11**, 1527–34.

Alves de Araújo, I. (1987). *O litoral Portugúes*. Serviço Nacional de Parques, Reservas e Conservação da Natureza, Lisbon.

Amaral Franco, J. do (1971–84). *Nova flora de Portugal (Continente e Açores)*, 2 volumes. Lisbon.

Amaral Franco, J. do and da Rocha Afonso, M. da Luz (1982). *Distribuição de pteridófitos e gimnospérmicas em Portugal*. Serviço de Parques, Reservas e Património Paisagistico, Lisbon.

Amorim, S. A. (1989). *A quarter of a century commitment with nature and technology*. Amorim, Lourosa.

Anderson, A. W. (1956). *Plants of the Bible*. Crosby Lockwood, London.

Anderson, B. and Anderson, E. (1991). *Landscapes of Portugal (Algarve). A countryside guide*. Sunflower Books, London.

Angeles Mendiola, M. (1989). *Plantas aromaticas de la Espāna Peninsular*. Ediciones Mundi-Prensa, Madrid.

Anon. (1850). *Da fabricação de águas-ardentes e de liquores de diversas fructas*. A. J. de Paula, Lisbon.

Anon. (1898). *Exposição de alfaia agrícola na Real Tapada da Ajuda em 1898: Documentos*. Imprensa Nacional, Lisbon.

Anon. (1954). Protecção à natureza em Portugal no anno de 1815. *Prot. Natur.*, 3, IX–X.

Anon. (1959). *Cartas agricola e floresta de Portugal, 566–612 (Algarve, 46 mapas), escala 1:25 000*. Secretaria de Estado da Agricultura, Lisbon.

Anon. (1964). Eucalipto: *Eucalyptus globulus* L. Ficha de características. *Folhosas Portuguesas*. MOP–Laboratório Nacional de Engenharia Civil, Lisbon.

Anon. (1973*a*). *Mapas do Algarve, Ortofotomapa, Escala 1:10 000*. Insituto Geográfico e Cadastral, Lisbon.

Anon. (1973*b*). Primeiro inventário das áreas ou zonas a proteger em Portugal continental. *Prot. Natur.*, n. s. **14**, 38–40.

Anon. (1977–9). *Cartas militar de Portugal, 577–612, Escala 1:250 000, Série M 888*, 2nd edn. Serviço Cartográfico de Exército, Lisbon.

Anon. (1980). *Guia dos produtos fitofarmacêuticos*. Vol. I *Insecticidas e fungicidas*, Vol. II *Herbicidas*. Ministério da Agricultura e Pescas, Lisbon.

Anon. (1981). Spanish citrus crop. *Financial Times*, 15 January 1981, London.

Anon. (1982*a*). *Documentação geográfica militar, Série M586 (Folhas 7/8)*. Serviço Cartográfico de Exército, Lisbon.

Anon. (1982*b*). *Sistemas de referenciação*. Serviço Cartográfico de Exército, Lisbon.

Anon. (1983). *Plantas aromáticas: óleos essenciais. I^{as} jornadas nacionais de plantas aromáticas e óleos essenciais.* Universidade de Coimbra.

Anon. (1984). *Produtos com venda autorizada. Suplemento e apéndices I–XXI.* Ministério da Agricultura e Pescas, Lisbon.

Anon. (1986). *Os sistemas lagunares do Algarve.* Universidade do Algarve, Faro.

Anon. (1987*a*). Grape harvest nears end. *Algarve News,* 25 September 1987.

Anon. (1987*b*). *1° Congresso Áreas Protegidas, Comunicaões e Conclusões.* Serviço Nacional de Parques, Reservas e Conservação da Natureza. Lisbon.

Anon. (1988*a*). *Estatísticas agricolas, Continente, Açores e Madeira 1987.* Instituto Nacional de Estatística. Lisbon.

Anon. (1988*b*). *Atlas de Portugal.* Reader's Digest (Portugal), Lisbon.

Anon. (1989*a*). *O eucalipto em Portugal.* Instituto Nacional do Ambiente, Lisbon.

Anon. (1989*b*). *Estatísticas demográficas, Continente, Açores e Madeira 1988.* Instituto Nacional de Estatística, Lisbon.

Anon. (1989*c*). *Estatísticas do turismo, Continente, Açores e Madeira 1988.* Instituto Nacional de Estatística, Lisbon.

Anon. (1989*d*). Soporcel : relatório e contos 1989. *Diário de Notícias,* 28 May 1990, pp. 14–16. Lisbon.

Anon. (1989*e*). *Mapa, Projecto Corine — Biótopos: 4. Inventário dos sitios de interesse para a conservação, Escala 1:250 000.* Secretaria de Estado do Ambiente e Recursos Naturais, Serviço Nacional de Parques, Reservas e Conservação da Natureza, Lisbon.

Anon. (1989*f*). *Mapa, Projecto Corine — Altimetria — Hidrografia — Áreas protegidas.* Secretaria de Estado do Ambiente e Recursos Naturais, Serviço Nacional de Parques, Reservas e Conservação da Natureza, Lisbon.

Anon. (1989*g*). *Mapa, Projecto Corine — Hidrografica — Biótopos — Áreas protegidas.* Secretaria de Estado do Ambiente e Recursos Naturais, Serviço Nacional de Parques, Reservas e Conservação da Natureza, Lisbon.

Anon. (1989*h*). Oídio da videira. *Agricultura 92,* **8,** 20–4.

Anon. (1989*i*). *Aquaculture: l'émergence d'une industrie nouvelle.* Organisation de Cooperation et de Développement Économiques, Paris.

Anon. (1990*a*). *All about Algarve, Vacation 90.* Região de Turismo do Algarve, Faro.

Anon. (1990*b*). *Aeroporto de Faro.* Aeroportos e Navegação Aéreo, Faro.

Anon. (1990*c*). Silvicultura: reforçada política de reflorestação. *Agricultura 92,* **20,** 31–8.

Anon. (1990*d*). *Moçárabe em peregrinação a S. Vicente (de Mértola ao Cabo de S. Vicente).* Integrada no projecto (sete itinerários medievais). Prod. Caminus: Commission of the European Communities, Brussels.

Anon. (1990*e*). Mão-de-obra agrícola está muito envelhecida. *Expresso,* 3 de Fevreiro 1990. Lisbon.

Anon. (1990*f*). Turismo cinegético: 113 zonas num novo modelo. *Agricultura 92,* **20,** 48–50.

Anon. (1990*g*). Brazilians thrive as Florida oranges feel chill. *Financial Times,* 11 July 1990. London.

Anon. (1990*h*). Inefficient agriculture struggles to become competitive. *Financial Times,* 24 October 1990. London.

Anon. (1990*i*). Authorities seek Algarve alternatives. *Financial Times,* 24 October 1990. London.

Anon. (1990*j*). The New Iberia. *Economist,* 5 May 1990. London.

Anon. (1990*k*). Tourism vs. ecology: the final cost. *Algarve Mag.,* January 1990, pp. 24–8.

Anon. (1990*l*). The pillage of Portugal. *Algarve Mag.,* February 1990, pp. 38–40.

Anon. (1990*m*). O sucesso das novas tecnologias genéticas florestais. *Agricultura 92,* **21,** 28–9.

Anon. (1990*n*). *Portugal estâncias termais: guia oficial. 1990/1*. Direcção-Geral do Turismo, Direcção-Geral de Geologia e Minas, Lisbon.

Anon. (1990*o*). *The environmental program for the Mediterranean. Preserving a shared heritage and managing a common resource*. The World Bank, New York and the European Investment Bank, Luxembourg.

Anon. (1990*p*). *Forest product prices 1969–1988*. FAO Forestry Paper 95. Food and Agricultural Organization of the United Nations. Rome.

Anon. (1991). *Aves de Portugal: birds of Portugal*. Serviços Nacional de Parques Reservas e Conservação da Natureza, Lisbon.

Ataíde Oliveira, F. X. (1908). *Monografia de Vila Real de Santo António*. Livraria Figueirinhas, Oporto. (Reprinted, Algarve em Foco, Faro).

Ataide Oliveira, F. X. (1913). *Monografia da Luz de Tavira*. Empresa Gráfica 'A Universal', Oporto.

Ataide Oliveira, F. X. (1914). *Monografia de Estoi (A vetusta Ossonobra)*. Comp. Port. Oporto.

Ataide Oliveira, F. X. (1986). *Monografia do Concelho de Olhão*, reprint of 1906 edn. Algarve em Foco, Faro.

Ataide Oliveira, F. X. (1987*a*). *Monografia de Porches*, reprint of 1912 edn. Algarve em Foco, Faro.

Ataide Oliveira, F. X. (1987*b*). *Monografia de São Bartolomeu de Messines*, reprint of 1909 edn. Algarve em Foco, Faro.

Ataide Oliveira, F. X. (1987*c*). *Monografia de Estombar*, reprint of 1911 edn. Algarve em Foco, Faro.

Ataide Oliveira, F. X. (1987*d*). *Monografia do Algoz*, reprint of 1905 edn. Algarve em Foco, Faro.

Ataide Oliveira, F. X. (1989*a*). *Monografia do Concelho de Loulé*, 3rd edn. Algarve em Foco, Faro.

Ataide Oliveira, F. X. (1989*b*). *Monografia de Paderne*, reprint of 1910 edn. Algarve em Foco, Faro.

Ataide Oliveira, F. X. (n.d.). *Monografia de Alvôr*, reprint of 1907 edn. Algarve em Foco, Faro.

Azeredo, A. C., Cabral, C., and Freire da Andrade, C. (1969, 1984). Aspectos catamórficos e anamórficos do litoral do Algarve (Portugal). (*Correio do Sul*), Lisbon, and pp. 225–39 in *LPN Boletim* N°. 18, 1984, **I**, Lisbon.

Backeberg, C. (1976). *Cactus Lexicon: enumeratio diagnostica Cactacearum*. Blandford Press, Poole.

Baedeker, K. (1898). *Spain and Portugal. Handbook for travellers*. Baedeker, Leipzig.

Baedeker, K. (1908). *Spain and Portugal. Handbook for travellers,* 3rd edn. Baedeker, Leipzig.

Barbero, M., Bonin, G., Loisel, R., and Quézel, P. (1990). Changes and disturbances of forest ecosystems caused by human activities in the western part of the Mediterranean basin. *Vegetatio*, **87**, 151–73.

Barranco, D. and Rallo, L. (1984). *Las variedades de olivo cultivadas en Andalucia*. Junta de Andalucia: Consejeria de Agricultura, Pesca y Alimentacion Córdoba.

Barros, L., Graça, J., and Pereira, H. (1986). A produção e a transformação da cortiça no Algarve. *4° Congresso do Algarve. Textos das comunicacões*, Vol. II, pp. 1061–6. Racal Clube, Silves.

Baumann, H. (1984). *Le bouquet d'Athéna. Les plantes dans la mythologie et l'art Grec* (translated from the French). Flammarion, Paris.

Beato, M. de F. (1982). *Catálogo das publicações 1865–1981*. Serviços Geológicos de Portugal, Lisbon.

Beato, M. de F. (1986). *Catálogo das publicações adicional*. Serviços Geológicos de Portugal, Lisbon.

Beckett, E. (1988). *Wild flowers of Majorca, Minorca, Ibiza: with keys to the flora of the Balearic Islands*. Balkema, Rotterdam.

Bivar Gomes da Costa Weinholz, M. (1883). *Dissertação sobre a cultura da figueira no Algarve*, unpublished. Instituto Superior de Agronomia, Lisbon.

Boardman, J. (1976). The olive in the Mediterranean: its culture and use. *Philos. Trans. R. Soc. London*, Ser. B, **278**, 187–96.

Bolton, M. (1989). Bird use of wetlands in the western Algarve. *A Rocha Observatory Report 1988*, pp. 37–45. Portimão.

Bonnet, C. (1850). Mémoire sur le royaume de l'Algarve. *Mem. Acad. Real Sci. Lisbon* II, **2**, 1–176k.

Bordalo da Rocha, R. B. (1976). *Estudo estratigráfico e paleontológico do Jurássico do Algarve Ocidental*, 1st edn. Ciências da Terra, Universidade Nova de Lisboa, Lisbon.

Bordalo da Rocha, R., Ramalho, M. M., Manuppella, G., and Zbyszewski, G. (1979). *Cartas Geológica de Portugal. Notícias explicativas da folhas*. Serviços Geológicos de Portugal, Lisbon.

Borg-Karlson, A. -K., Bergström, G., and Groth, I. (1985). Chemical basis for the relationship between *Ophrys* orchids and their pollinators I. *Chem. Scripta*, **25**, 283–94.

Borg-Karlson, A. -K., Bergström, G. and Kullenberg, B. (1987). Chemical basis for the relationship between *Ophrys* orchids and their pollinators II. *Chem. Scripta*, **27**, 303–11.

Botelho da Costa, J. (1985). *Caracterização e constituição do solo*, 3rd edn. Fundação Calouste Gulbenkian, Lisbon.

Bowman, F. T. (1956). *Citrus growing in Australia*. Angus and Robertson, Sydney.

Bramwell, D. and Bramwell, Z. (1974). *Wild flowers of the Canary Islands*. Thornes, London.

Bravo de Lima, M., Belchior, A. P., and Estabrook, G. F. (1988). Uniformity and constancy of wine tasters evaluating the same wines on two different occcasions. *Ciência Téc. Vitiv.*, **7**, 73–85.

Brazão Gonçalves, E. (1988). *Dictionário do falar Algarvio*. Região de Turismo do Algarve, Lisbon.

Buczacki, S. and Harris, K. (1981). *Collins guide to the pests, diseases and disorders of garden plants*. Collins, London.

Bugalho Semedo, C. M. (1988). *A intensificação da produção hortícola*, 3rd edn. Publicações Europa-América, Mem Martins.

Butzer, K. W., Mateu, J. F., Butzer, E. K., and Kraus, P. (1985). Irrigation agrosystems in eastern Spain: Roman or Islamic origins? *Ann. Assoc. Am. Geogr.*, **75**, 479–509.

Cabral, J. M. P., Herculano de Carvalho, A., and Lima, M. B. (1977). Aplicação de métodos de taxonomia numérica na classificação de águas minerais de Portugal Continental. *Comunic. Serv. Geol. Port.*, **61**, 343–63.

Cabral Rolo, J. A. and Seita Coelho, I. (1988). *A(s) agricultura(s) Algarvia(s)*, 2 volumes. Ministério do Planeamento e da Administração do Território — Comissão de Coordenação da Região do Algarve. Faro.

Cabreira, T. (1918). *O Algarve economico*. Imprensa Libanio da Silva, Lisbon.

Cabrita Neto, J. M. (1980). A importância das frutas secas e frescas na economia do Algarve. *1° Congresso Nacional sobre O Algarve. Textos das comunicações*. Racal Clube, Aldeia das Açoteias.

Campos Pereira, J. de (1915). *Economica e finanças: a proprieda de rústica em Portugal*. Imprensa Nacional, Lisbon.

Carneiro, L. C., and Bravo de Lima, M. (1987). Ampelographic characterization of grapevine varieties using leaf shape. *Ciência Téc. Vitiv.*, **6**, 67–78.

Carvalho de Vasconcelos, J. M. (1988). *Seminário Corine Projecto Biótopos. II Congresso Corine*. SNPRCN, Lisbon.

Carvalho Quintela, A. de, Cardoso, J. M., and Mascarenhas, J. M. (1987). *Aproveitamentos hidráulicos Romanos a sul do Tejo*. Direcção-Geral dos Recursos e Aproveitamentos Hidráulicos. SNPRCN, Lisbon.

Casadinho Parrinha Duarte Regato, M. A. (1989). A agricultura biológica no Parque Natural da Ria Formosa. *Comm. II Congresso de áreas protegidas: Fundação Calouste Gulbenkian, Lisboa, 4–8 de Dezembro de 1989*, pp. 59–63. SNPRCN, Lisbon.

Casimiro Martins, J, Conceição Gonçalves, M., Menezes Sequeira, E. and Pereira Gomes, M. (1992a). A qualidade do solo e a importância do recurso-terra no Algarve. *7° Congresso do Algarve. Textos das comunicações*: 359–64. Racal Clube, Vilamoura.

Casimiro Martins, J, Conceição Gonçalves, M., Menezes Sequeira, E. and Pereira Gomes, M. (1992b). Availiação de erodibilidade e dos riscos de erosão do solo no Algarve. *7° Congresso do Algarve. Textos das comunicações:* 413–18. Racal Clube, Vilamoura.

Castro Henriques, P., Cabrita, A., and Cunha, R. (1990). *Parques e reservas naturais de Portugal*. Verbo, Lisbon.

Castroviejo, S., Lainz, M., López González, G., Montserrat, P., Muñoz Garmendia, F., Paiva, J., and Villar, L. (eds.) (1986–90). *Flora Iberica*, 2 volumes. Real Jardín Botánico CSIC, Madrid.

Catela Pais Mousinho, D. M. (1984). Frutos secos no Algarve. Que futuro?. *3° Congresso do Algarve. Textos das comunicações*. Vol. 2, pp. 1009–1017. Racal Clube, Silves.

Cavaco, C. (1976). *O Algarve Oriental, as vilas, o campo e o mar*, 2 volumes. Gabinete do Planeamento da Região do Algarve, Faro.

Cavaco, C. (1983). *A agricultura do Algarve, segundo o recenseamento agricola de 1979*. Centro de Estudos Geográficos, Instituto Nacional de Investigação Científica, Universidade de Lisboa, Lisbon.

Chester, D. K. and James, P. A. (1989). A preliminary report on the Quaternary landforms of three valleys in the Algarve Region, Southern Portugal. *Liverpool Pap. Geog.*, **3**. University of Liverpool.

Chiej, R. (1984). *The Macdonald encyclopedia of medicinal plants*. Macdonald, London.

Chittenden, F. J. (1951). *The Royal Horticultural Society dictionary of gardening*, 4 volumes. Oxford University Press.

Clutton-Brock, J. (1987). *The natural history of domesticated mammals in early times*. Cambridge University Press.

Cobianchi, D., Bergamini, A., and Cortesi, A. (1989). *El ciruelo*. Ediciones Mundi-Prensa, Madrid.

Colino Sueiras, J., Bello Fernandez, E., Carreño Sandoval, F., Lopez Martinez, M., Noguera Mendez, P., and Riquelme Perea, F. (1990). *Precios, productividad y renta en las agriculturas Españolas*. Ediciones Mundi-Prensa, Madrid.

Collenette, S. (1988). The sweet olives of Saudi Arabia. *Kew Mag.*, **5**, 36–8.

Comissão de Coordenação da Região do Algarve (1985). *Programa de desenvolvimento regional. Algarve 1986–90*. Comissão de Coordenação da Região do Algarve, Faro.

Condit, I. J. (1925). Fig industry of Portugal. *Calif. Fruit News*, **71** (1918), 4, 9.

Condit, I.J. (1947). *The fig*. Chronica Botanica. Waltham, Mass.

Cooke, G. B. (1961). *Cork and the cork tree*. Pergamon. Oxford.

Cornaert, M. H., Steenmans, C., Henriques, R. G., and Mariette, V. (1988). *Workshop on Innovative Services for IGIS users: Symposium on Alpine and Mediterranean areas: a challenge for remote sensing: the E. C. CORINE Land Cover Data Base: an aid for European and national environmental planning policies. 15–20 May 1988. Capri.* Commission of the European Communities, Brussels.

Corrêa da Mello Leotte, F. (1901). *Arboricultura Algarvia, figueira, amendoeira e alfarrobeira*. Adolpho de Medonça, Lisbon.

Correia, A. D. A. (1982). *Fisiografia e geografia física da província do Algarve.* Universidade Técnica de Lisboa, Instituto Superior de Agronomia, Lisbon.

Correia, E. (1983). *Alguns apontamentos sobre o concelho de Aljezur,* 2nd edn. Câmara Municipal de Aljezur.

Correia Peixoto, L. (1991). *Apontamentos para a história da pesca sardinha e da construção naval em Peniche.* Câmara Municipal de Peniche.

Costa, A. da and Franquinho, L. O. (1986). *Plantes e flores Madeira,* 8th edn. Francisco Ribeiro, Funchal.

Costa, J. C. and Lousã, M. (1986). *Estudo e cartografia da vegetação da Reserva Natural da Ria Formosa.* Instituto Superior de Agronomia, Lisbon.

Costa, M. G. and Entrudo Fernandes, J. M. (1990). *Frankliniella occidentalis (tripe da California): nova praga de grande importância no Algarve.* Direcção Regional de Agricultura do Algarve, Faro.

Costa Guedes, L. da (1988). *Aspectos do Reino do Algarve nos seculos XVI e XVII.* Arquivo Histórico Militar, Lisbon.

Costa Primo, S. da (1936). Quelques observations sur la végétation de Sagres et du Cap de S. Vicente. *Bull. Soc. Port. Sci.,* **12,** 125–39.

Coutinho, A. X. P. (1915). *Notas da Flora de Portugal.* Livrarias Aillaud e Bertrand, Paris-Lisbon.

Coutinho, A. X. P. (1939). *Flora de Portugal (Plantes vasculares),* 2nd edn. Bertrand, Lisbon.

Crawford, R. M. M. (1989). *Studies in plant survival.* Blackwell, Oxford.

Crespo, E. G. and Oliveira, M. E. (1989). *Atlas da distribuição dos anfibios e répteis de Portugal Continental.* Serviço Nacional de Parques, Reservas e Conservação da Natureza, Lisbon.

Cruz, C. S. (1986). Estruturas halofíticas do Algarve (algumas considerações sobre a vegetação das arribas areais litorais e dos sapais). *4° Congresso do Algarve. Textos das comunicacões,* Vol. 2, pp. 727–34. Racal Clube, Silves.

Cruz Rosa, M. da (1978). *Estudo da produção de aguardente da medronho no Algarve.* Universidade Técnica de Lisboa, Tavira.

Cunha Mota, M. da (1919). *Catálogo geral de plantas, sementes e outros artigos.* A Intermediaria, Oporto.

Curvelo-Garcia, A. S., Bravo de Lima, M., Spranger-Garcia, M. I., and Coelho, D. (1987). Caracterização analítica de vinhos rosados por aplicação das técnicas de taxonomia numérica. *Ciência Téc. Vitiv.,* **6,** 79–97.

Cyrén, O. (1930–1). Vegetationsbilder aus Portugal. *Vegetationsbilder,* Vol. 3–4, pp. 21. Fischer, Jena.

D'Almeida, A. and De Almeida, J. (1966). *Inventário hidrológico de Portugal (Algarve).* Instituto de Hidrológia da Lisboa, Lisbon.

D'Almeida de Bivar Weinholz, F. (1891). *Memória sobre a economia rural do 9ª Região.* Instituto de Agronomia, Tavira.

Damásio, M. I. (1990). O castanheiro. *Agricultura 92,* **21,** 30–2.

Daveau, J. (1897). La flore littorale du Portugal. *Bol. Soc. Brot.,* **14,** 3–54.

Daveau, S. (1977). Répartition et rythme des précipitations au Portugal. *Memórias do Centro de Estudos Geográficos,* **3.** Lisbon.

Daveau, S. (1988). Progressos recentes no conhecimento da evolução holocénica da cobertura vegetal, em Portugal e nas regiões vizinhas. *Finisterra,* **45,** 101–52.

Daveau, S. e colaboradores (1985). Mapas climáticos de Portugal. Nevoeiro e nebulosidade. *Memórias do Centro de Estudos Geográficos,* **7.** Lisbon.

Davidson, L. and Jeppe, B. (1981). *Acacias: a field guide to the acacias of southern Africa.* Centaur, Johannesburg.

Davies, P., Davies, J., and Huxley, A. (1988). *Wild orchids of Britain & Europe*. Hogarth Press, London.

Davis, S. O., Droop, S. J. M., Gregerson, P., Henson, L., Leon, C. J., Villa-Lobos, J. L., Synge, H., and Zantunska, J. (1986). *Plants in danger. What do we know?* International Union for the Conservation of Nature and Natural Resources, Gland, Switzerland.

Debazac, E. F. (1983). Temperate broad-leaved evergreen forests of the Mediterranean region and Middle East. In *Temperate broad-leaved evergreen forests* (ed. J. D. Ovington), pp. 107–23. Elsevier, Amsterdam.

Debussche, M. and Isenmann, P. (1990). Introduced and cultivated fleshy-fruited plants: consequences for a mutualistic Mediterranean plant–bird system. In *Biological invasions in Europe and the Mediterranean Basin* (ed. F. di Castri, A. J. Hansen, and M. Debussche), pp. 399–416. Kluwer, Dordrecht.

Devereux, C. M. (1982). Climate speeds erosion of the Algarve's valleys. *Geogr. Mag.*, **53**, 10–17.

Dias, A. S., Dias, L. S., and Pereira, I. P. (1989). *Cistus ladanifer* L. esteva. Caracterização e perspectivas de utilização. *Comm. II Congresso de áreas protegidas: Fundação Calouste Gulbenkian, Lisboa, 4–8 de Dezembro de 1989*, pp. 275–83. SNPRCN, Lisbon.

Dias, M. H. P. (1986). A flora e a vegetação do Algarve achegas para o seu conhecimento e protecção. *4° Congresso do Algarve. Textos das comunicações*, Vol. 2, pp. 705–10. Racal Clube, Silves.

Dray, A. M. (1985). *Plantas a proteger em Portugal Continental*. Serviço Nacional de Parques, Reservas e Conservação da Natureza, Lisbon.

Duarte, C., Lousâ, M., and Moreira, E. I. (1984). A vegetação ribeirinha do Sotavento Algarvio. *3° Congresso sobre o Algarve. Textos das comunicações*, Vol. 1, pp. 513–21. Racal Clube, Silves.

Duarte Amaral, J. (1982). *Os citrinos*, 3rd edn. Teixeira & Cª, Lisbon.

Eliseu, J. A. (1982). *Mexilhoeira Grande. Monografia etnografia*. J. A. Eliseu.

Emilia Rosa, M. and Amaral Fortes, M. (1987). Efeito do gás contido nas células na compressão da cortiça. *Bol. Inst. Flor.*, **590**, 331–6.

Emilia Rosa, M. and Amaral Fortes, M. (1988*a*). Temperature-induced alterations of the structure and mechanical properties of cork. *Mater. Sci. Eng.*, **100**, 69–78.

Emilia Rosa, M. and Amaral Fortes, M. (1988*b*). Thermogravimetric analysis of cork. *J. Mater. Sci. Lett.*, **7**, 1064–5.

Emilia Rosa, M. and Amaral Fortes, M. (1988*c*). A estrutura celular da cortiça: leis estatísticas de formação das células. *Técnica*, September 1988, pp. 3–13. Lisbon.

Emilia Rosa, M. and Amaral Fortes, M. (1988*d*). Densidade da cortiça: factores que a influênciam. *Bol. Inst. Prod. Flor. Cortiça*, **593**, 65–9.

Emilia Rosa, M. and Amaral Fortes, M. (1988*e*). Stress relaxation and creep of cork. *J. Mater. Sci.*, **23**, 35–42.

Emilia Rosa, M. and Amaral Fortes, M. (1988*f*). Rate effects on the compression and recovery of dimensions of cork. *J. Mater. Sci.*, **23**, 879–85.

Emilia Rosa, M. and Amaral Fortes, M. (1989). Effects of water vapour heating on structure and properties of cork. *Wood Sci. Technol.*, **23**, 27–34.

Entrudo Fernandes J. M. and Grilo, C. (1989). *O rato toupeira*. Direcção Regional de Agricultura do Algarve. Faro.

Espirito Santo, M. D. (1989*a*). *Flora das dunas e platformas arenosas do sudoeste meridional*. Instituto Superior de Agronomia, Lisbon.

Espirito Santo, M. D. (1989*b*). *Flora do litoral xistoso do sudoeste meridional*. Instituto Superior de Agronomia, Lisbon.

Espirito Santo, M. D. (1989*c*). *Flora do barrocal Algarvio ocidental (do Burgau ao Cabo de S. Vicente)*. Instituto Superior de Agronomia, Lisbon.

Espirito Santo, M. D. (1989*d*). *Contribuição para o conhecimento da flora dos estevais do sudoeste meridional.* Instituto Superior de Agronomia, Lisbon.

Espirito Santo, M. D. (1989*e*). *Flora da Serra de Espinhaço de Cão.* Instituto Superior de Agronomia, Lisbon.

Espirito Santo, M. D. (1990). *Fitocenoses de àguas salobras dos concelhos de Aljezur e Vila do Bispo.* Instituto Superior de Agronomia, Lisbon.

Estácio da Veiga Silva Pereira, M. L. (1866). Plantas de Serra de Monchique em 1866. *J. Sci. Math.*, **1866–96**, 120–213.

Fabião, A. M. D. (1987). *Árvores e florestas.* Publicações Europa-America, Mem Martins.

Faleiro Silva, P. R. (1989). A inovação tecnológica na salicultura. *Comm. II Congresso de áreas protegidas: Fundação Calouste Gulbenkian, Lisboa, 4–8 de Dezembro de 1989*, pp. 65–72. SNPRCN, Lisbon.

Faustino Crespo, M. E. (1968). *A economica marítima do Algarve na segunda metade do século XVIII.* Faculdade de Letras de Lisboa, Lisbon.

Feio, M. (1983). *Le bas Alentejo et l'Algarve.* Instituto Nacional de Investigação Científica, Centro de Ecologia Aplicada da Universidade de Évora, Évora.

Feio, M. (1989). *A reconversão da agricultura e a problemática do eucalipto.* Associação Central de Agricultura Portuguesa, Lisbon.

Ferguson, L., Sakovich, N., and Roose, M. (1990). California citrus rootstocks. *Div. Agric. Nat. Res.*, **21477**. University of California, Oakland, CA.

Ferreira, E. L. S. (1990). Visita de estudo a Espanha relaçionada com a cultura de nespereira (Callosa d'Ensarria — Espanha) 30 April–5 Maio 1989. *Estudos Técnicas*, **16**. Direcção Regional de Agricultura do Algarve, Faro.

Ferreira de Mira, P. (1939). *Valor alimentar e terapêutico do azeite.* Casa Portuguesa, Lisbon.

Ferro, J. P. (1990). *Para a história de Lagôa no Século XVIII: a Criação do concelho (1773).* Algarve em Foco, Faro.

Fideghelli, C. (1987). *El melocotonero.* Ediciones Mundi-Prensa, Madrid.

Figueiredo e Meneses da G. do Espírito Santo, A. de (1970). *Contribuição para um estudo económico da produção de uva de mesa na região extra-tempora do Algarve,* **RF-1778**. Instituto Superior de Agronomia, Universidade Técnica de Lisboa.

Flores Dominguez, A. (1990). *La higuera: frutal Mediterráneo para climas cálidos.* Ediciones Mundi-Prensa, Madrid.

Folques, J. (1987). Notas sobre erosão superficial. **ICT: ITG 5**. *Informacão Técnica Geotecnia.* Laboratório Nacional de Engenharia Civil. Lisbon.

Folving, S. and Mégier, J. (1989). *Remote sensing in the management of less favoured areas: European Collaborative Programme. Land use planning for the European marginal areas.* Joint Research Centre, Commission of the European Communities. **SP.1.89.22**. JRC-Ispra.

Fonseca Leal de Oliveira, A. da (1990). Subsidos para a delimitação do Parque Natural de Monchique. *6° Congresso do Algarve: textos das comunicações*, Vol. 2, pp. 487–95. Racal Clube, Silves.

Ford, E. B. and Placito, P. J. *Fogous*, (1989). MS Bodleian Library and All Souls College, University of Oxford.

Fritz, E. S., Barahona, M. H., Santos Oliveira, J., Esteves Duarte, F., Toscano, F., and Abecassis Empis, J. (1988). *Potencialidades da aquacultura em Portugal.* Fundação Luso-Americana para o Desenvolvimento, Lisbon.

Furtado, P. (1990). Nogueira: a cultura da nogueira no Algarve: estudo da importância da *Cydia pomonella* L. como factor limitante. *Experimentação*, **72**. Direcção Regional de Agricultura do Algarve, Faro.

Gaillard, J. -P. (1987). *L'avocatier. Sa culture, ses produits.* Maisonneuve et Larose, Paris.

Galan Sauco, V. (1990). *Los frutales tropicales en los subtropicos: I. Aguacate–mango–litchi y longan.* Ediciones Mundi-Prensa, Madrid.

Galego, J. and Daveau, S. (1986). O numerante de 1527–1532, Tratamento cartográfico. *Mem. Cent. Ets. Geogr.*, **9**. Lisbon.

Galil, J. and Neeman, G. (1977). Pollen transfer and pollination in the common fig *(Ficus carica* L.). *New Phytol.*, **79**, 163–71.

Garcia, J. M. and Cunha, R. (1990). *Sagres*. Câmara Municipal de Vila do Bispo.

Garcia Alonso, C. R. (1990). *El ajo: cultivo y aprovechamiento*. Mundi-Prensa, Madrid.

Garcia Rivas, I. (1990). Cultivo del litchi en la costa Mediterranea. *Hojas Divulgadores*, **4**, 2–23.

Gardé, A. and Gardé, N. (1988). *Cultivas hortícolas*, 6th edn. Clássica Editora, Lisbon.

Gascon, J. A. G. (1955). *Sudsídios para a monografia de Monchique*. Guerreiro Gascon, Portimão.

Gaspar, J. (1981). *Portugal em mapas e em números*, 2nd edn. Livros Horizonte, Lisbon.

Gimeno, E. Q. (1987). *El cultivo moderno del almendro*. Editorial de Vecchi, Barcelona.

Goes, E. and da Silva, A. A. (1986). Árvores monumentais do Algarve. *4° Congresso do Algarve. Textos das comunicações*, Vol. 2, pp. 719–25. Racal Clube, Silves.

Goldblatt. P. (1980). Systematics of *Gynandriris* (Iridaceae), a Mediterranean–southern African disjunct. *Bot. Not.*, **133**, 239–60.

Gomes Guerreiro, M. (1977). *O Algarve do futuro na perspectiva ecológica*. Secretaria de Estado do Ambiente, Lisbon.

Gómez-Campo, C. and Malato Beliz, J. (1985). The Iberian peninsula. In *Plant conservation in the Mediterranean area (Geobotany*, **7**, ed. C. Gómez-Campo), pp. 47–70. Junk, Dordrecht.

Gómez-Campo, C., Bermúdez-de-Castro, L., Cagiga, M. J., and Sánchez-Yélamo, M. B. (1984). Endemism in the Iberian Peninsula and Balearic Islands. *Webbia*, **38**, 709–14.

Goncalves Vieira, J. (1911). *Memoria monographica de Villa Nova de Portimão*. Tipografia Universal, Oporto.

Graça Silva, M. da. (1989). Parque Natural da Ria Formosa: interesse botânico. *Comm. II Congresso de áreas protegidas: Fundação Calouste Gulbenkian, Lisboa, 4–8 de Dezembro de 1989*, pp. 579–87. SNPRCN, Lisbon.

Graf, A. B. (1978). *Exotica Series 3: pictorial cyclopedia of exotic plants from tropical and near-tropic regions*, 9th edn. Roehrs, Fairfield, New Jersey.

Grandjot, W. (1965). *Schroeder Reiseführer durch das Pflanzenreich der Mittelmeerländer*, 3rd edn. Kurt Schroeder, Bonn.

Grasselly, C. and Crossa-Raynaud, P. (1980). *L'amandier*. Maisonneuve et Larose, Paris.

Green, P. S. and Kupicha, F. K. (1979). Notes on the genus *Olea*. *Kew Bull.*, **34**, 69–75.

Green, P. S. and Wickens, G. E. (1989). The *Olea europea* complex. In *Davis and Hedge Festschrift* (ed. K. Tan), pp. 287–299. Edinburgh University Press.

Guerreiro, F. (1988). *Pequena monografia de Pechão*. Algarve em Foco, Faro.

Guerreiro Costa, J. M. (1977). *Caracterização geral da agricultura Algarvio. RB-6247*. Universidade Técnica de Lisboa (Instituto Superior de Agronomia), Lisbon.

Guerrero, A. (1991). *Nueva olivicultura*, 2nd edn. Ediciones Mundi-Prensa, Madrid.

Guilaine, J. (1979). The earliest Neolithic in the West Mediterranean. *Antiquity*, **53**, 22–9.

Guillerm, J. L., le Floc'h, E., Maillet, J., and Boulet, C. (1990). The invading weeds within the Western Mediterranean Basin. In *Biological invasions in Europe and the Mediterranean Basin* (ed. F. di Castri, Hansen, A. J. and Debussche, M.). Kluwer, Dordrecht.

Hadlington, S. (1990). Unlocking secrets of the vine. *Chem. Br.*, **26**, 1134.

Hartmann, H. T., Opitz, K. W., and Beutel, J. A. (1980). Olive production in California. *Div. Agric. Soc.*, **2474**. University of California, Oakland.

Heitor, N. and Fernandes, P. (1989). *Guia de vinhos Portugueses 1990*. Barbosa e Almeida, Lisbon.

Henriques, J. A. (1890). Exploração botanica em Portugal por Tournefort em 1689. *Bol. Soc. Brot.*, **8**, 191–261.

Herklots, G. A. C. (1972). *Vegetables in south-east Asia*. George Allen and Unwin, London.

Herklots, G. A. C. (1976). *Flowering tropical climbers*. Dawson: Science History Publications, Folkestone.

Herrera, C. M. (1981). Fruit variation and competition for dispersers in natural populations of *Smilax aspera. Oikos*, **36**, 51–8.

Herrera, C. M. (1982). Seasonal variation in the quality of fruits and diffuse co-evolution between plants and avian dispersers. *Ecology*, **63**, 773–84.

Herrera, C. M. (1984). Avian interference of insect frugivory: an exploration into the plant–bird–fruit pest evolutionary triad. *Oikos*, **42**, 203–10.

Herrera, C. M. (1985). Predispersal reproductive biology of female *Osyris quadripartita* (Santalaceae), a hemiparasitic dioecious shrub of Mediterranean scrublands. *Bot. J. Linn. Soc.*, **90**, 113–27.

Herrera, C. M. (1986). Vertebrate-dispersed plants: why they don't behave the way they should. In *Frugivores and seed dispersal* (ed. A. Estrada and T. H. Fleming) pp. 5–18. Junk, Dordrecht.

Herrera, C. M. (1987). Vertebrate-dispersed plants of the Iberian Peninsula: a study of fruit characteristics. *Ecol. Monog.* **57**, 305–31.

Herrera, C. M. (1988a). Habitat-shaping, host plant use by a hemiparasitic shrub, and the importance of gut-fellows. *Oikos*, **51**, 383–6.

Herrera, C. M. (1988b). Plant size, spacing patterns, and host-plant selection in *Osyris quadripartita*, a hemiparasitic dioecious shrub. *J. Ecol.*,**76**, 995–1006.

Herrera, C. M. (1988c). The fruiting ecology of *Osyris quadripartita*: individual variation and evolutionary potential. *Ecology*, **69**, 233–49.

Herrera, C. M. (1989). Frugivory and seed dispersal by carnivorous mammals, and associated fruit characteristics, in undisturbed Mediterranean habitats. *Oikos*, **55**, 250–62.

Hills, L. D. (1980). The cultivation of the carob tree (*Ceratonia siliqua*). *Int. Tree Crops J.*, **1**, 27–47.

Horner, O. (1990). Medronho making. *Discover Portimão*, March 1990, pp. 6, 16, 27.

Ibar Albiñana, L. (1985). *Cultivo moderno del almendro*. Aedos, Barcelona.

Ibar Albiñana, L. (1986). *Cultivo del aguacate, chirimoyo, mango, papaya*, 3rd edn., Biblio. Agric. Aedos, Barcelona.

Iglesia Gonzalez, J. A. (1988). *La feijoa*. Mundi-Prensa, Madrid.

Ingram, J. (1989). Unpublished MS undergraduate project. Department of Plant Sciences, University of Oxford.

Innes, C. (1987). *Wild flowers of Spain*, 3 volumes. Cockatrice, Whitchurch.

Instituto Geográfico e Cadastral (1964). *Cartas corográfica de Portugal*. Série M 7810, 2nd edn. Lisbon.

Instituto Nacional de Investigação Científica (1989). *O Algarve na perspectiva da antropologia ecológica*. Lisbon.

Izhaki, I. and Safriel, U. N. (1989). Why are there so few frugivorous birds? Experiments in fruit digestibility. *Oikos*, **54**, 23–32.

Jacobsen, H. (1960). *A handbook of succulent plants: descriptions, synonyms and cultural details for succulents other than Cactaceae*, 3 volumes, English edn. Blandford Press, London.

Jones, M. P. (1980). Epiphytic macrolichens of the Algarve, Portugal. *Lichenologist*, **12**, 253–75.

Jones, R. (1978). *Wild flowers of the western Mediterranean*. Jarrold, Norwich.

Jordano, P. (1982). Migrant birds are the main seed dispersers of blackberries in southern Spain. *Oikos*, **38**, 183–93.

Jordano, P. (1989). Pre-dispersal biology of *Pistacia lentiscus* (Anacardiaceae): cumulative effects on seed removal by birds. *Oikos*, **55**, 375–86.

Juhren, M. C. (1966). Ecological observations on *Cistus* in the Mediterranean vegetation. *Forest Sci.*, **12**, 415–26.

Juniper, B. E., Robins, R. J., and Joel, D. M. (1989). *The carnivorous plants*. Academic Press, London.

Kardell, L., Steen, E., and Fabião, A. (1986). Eucalyptus in Portugal — threat or a promise? *Ambio*, **15**, 6–13.

Kjellberg, F., Gouyon, P. -H., Ibrahim, M., Raymond, M., and Valdeyron, G. (1987). The stability of the symbiosis between dioecious figs and their pollinators: a study of *F. carica* L. and *Blastophaga psenes* L. *Evol.*, **41**, 693–704.

Kopp, E., Sobral, M., Soares, T., and Woerner, M. (1989). *Os solos do Algarve e as suas características. Vista geral*. Ministério da Agricultura, Pescas e Alimentação, Direcção Regional de Agricultural do Algarve, Faro.

Kullenberg, B., and Bergström, G. (1976). *Hymenoptera Aculenta* males as pollinators of *Ophrys* orchids. *Zool. Scripta*, **5**, 13–23.

Landmann, G. (1818). *Historical, military, and picturesque observations on Portugal, illustrated by 75 coloured plates, including authentic plans on the sieges and battles fought in the peninsula in the late war*. Cadell and Davies, London.

Lanza, P. and Pizzetti, M. (1979). *Guia de arboles*. Grijalbo, Barcelona.

LaRue, J. H., Copeland, R. D., and Pehrson, J. (1983). Growing avocados in the San Joaquin Valley. *Div. Agric. Sci.*, **2904**. University of California, Berkeley.

Latymer, H. (1990). *The Mediterranean gardener*. Lincoln and Royal Botanic Gardens, Kew.

Lautensach, H. (1988). A cobertura vegetal. In *Geografia de Portugal II* (ed. O. Ribeiro, H. Lautensach, and S. Daveau), pp. 539–75. Sá da Costa, Lisbon.

Leach, H. M. (1982). On the origins of kitchen gardening in the ancient Middle East. *Garden Hist.*, **10**, 1–16.

Lebreton, P. and Rivera, D. (1988). Analyse du taxon *Juniperus phoenicea* L. sur les bases biochimiques et biométriques. *Natur. Monsp. Bot.*, **53**, 17–41.

Le Houérou, H. N. (1974). Fire and vegetation in the Mediterranean Basin. *Proceedings Annual Tall Timbers Fire Ecology Conference 1973*, pp. 237–77. Tall Timbers Research Station, Tallahassee, Florida.

Le Houérou, H. N. (1981). Impact of man and his animals on Mediterranean vegetation. In *Mediterranean-type shrublands. Ecosystems of the World 11*, (ed. F. di Castri, D. W. Goodall, and R. L. Specht), pp. 479–521. Elsevier, Amsterdam.

Le Houérou, H. N. (1987). Vegetation wild fires in the Mediterranean basin: evolution and trends. *Ecol. Medit.*, **13**(3), 13–24.

Leibold G. (1980). *Guia das plantas medicinais*. Editorial Presença, Lisbon.

Lepierre, C. (1935). *A industria do sal em Portugal*. Universidade Técnica de Lisboa, Lisbon.

Lima Miranda, D. J. de (1974). *Nogueiras*. Atlântida Editora, Coimbra.

Link, H. F. (1801). *Travels in Portugal*. Longman & Rees, London.

Lonsdale, A. M. (1980). *Merchant adventurers in the East*. Longman, London.

Lopes Leitão, J. (1992). Disponibilidades hídricas das barragens Arade /Funcho. *7° Congresso do Algarve, Textos das comunicações:* 377–83. Racal Clube, Vilamoura.

Lopez Palazon, J. (1970). *El almendro en Baleares*. Instituto Nacional de Investigaciones Agronomicas, Madrid.

Loureiro, F. de (1984). *Uma jornado ao Alentejo e ao Algarve. (Texto do cronista, João Cascão)*. Livros Horizonte, Lisbon.

Louro, E. (1986). *O livro de Alportel*, 2nd edn. Câmara Municipal de São Brás de Alportel.

Lousã, M., Moreira, I., Carvalho, J. S., and do Rosário, L. P. (1986). Barragem de Odeleite — minimizaçao dos impactos bióticos. *4° Congresso do Algarve. Textos das comunicações*, Vol. 2, pp. 649–55. Racal Clube, Silves.

Loussert, R. and Brousse, G. (1978). *L'olivier*. Maisonneuve et Larose, Paris.

Luna Lorente, F. (1986). *Plantas aromáticas: cultura e destilação*. Agri-Gado, Lisbon.

Luna Lorente, F. (1990). *El nogal*, 2nd edn. Ministerio de Agricultura, Pesca y Alimentacion. Ediciones Mundi-Prensa, Madrid.

Luther, W. and Fiedler, K. (1976). *A field guide to the Mediterranean sea shore*. Collins, London.

Mabberley, D. J. (1984). A monograph of *Melia* in Asia and the Pacific. The history of White Cedar and Persian Lilac. *Gard. Bull. Singapore*, **37**, 49–64.

Mabberley, D. J. (1990). *The plant-book*. Reprint with corrections. Cambridge University Press.

Mabberley, D. J. (1991). *Tropical rain forest ecology*, 2nd edn. Blackie, Glasgow.

Macário Correia, J. (1984). *Niveis de desenvolvimento do Algarve*. Comissão de Coordenação Região do Algarve, Faro.

Macedo, F. W. and Sardinha, A. M. (1985). *Fogos florestais*. Vol. 1. Publicações Ciência e Vida, Lisbon.

McMurtrie, M. (1973). *Wild flowers of the Algarve*. M. McMurtrie.

McMurtrie, M. (1986). *More wild flowers of the Algarve*. M. McMurtrie.

Madeira, J. A. (1969, 1984). *O Algarve. Costa mundial do sol*. Soc. do Jornal de Faro (Correio do Sul), Lisbon and Liga para a Protecção da Natureza, Lisbon.

Madeira Lobo, J. (1977). *Fruticultura de hoje*. Livraria Luso-Espanhola, Lisbon.

Malato Beliz, J. (1982). *A Serra de Monchique: flora e vegetação*. Serviço Nacional de Parques, Reservas e Património Paisagístico, Lisbon.

Malato Beliz, J. (1986). *O barrocal Algarvio: Flora e vegetação da amendoeira (Loulé)*. Serviço Nacional de Parques, Reservas e Património Paisagístico, Lisbon.

Malato Beliz, J. (1990a). *A Serra de Portel: flora e vegetação*. Serviço Nacional de Parques, Reservas e Conservação, Lisbon.

Malato Beliz, J. (1990b). A cobertura vegetal primativa e actual entre Mértola e Sagres–S. Vicente. In Anon. 1990d, pp. 68–72.

Manupella, G., Rocha, R. B., Marques, B., and Ramalho, E. M. M. (1984). Cartografia Geológica do Algarve. *3° Congresso sobre o Algarve. Textos das comunicações*, Vol. 2, pp. 693–704. Racal Clube, Montechoro.

Margalef, R. (1985). *Western Mediterranean. Key environments*. Pergamon, Oxford.

Marota Borrego, J. V. and Lopez Galarza, S. (1988). *Producción de fresas y fresones*. Mundi-Prensa, Madrid.

Marques Ferreira, J. (1989). A conservação da natureza, as áreas protegidas e a agricultura. *Agricultura 92*, **8**, 6–10.

Marques Moreira, J. (1987). *Alguns aspectos de intervenção humana na evolução da paisagem da Ilha de S. Miguel (Açores)*. Serviço Nacional de Parques, Reservas e Conservação da Natureza, Lisbon.

Martin, I., Bennett, R. J., and Gregory, D. J. (1984). The thirsty Algarve. *Geogr. Mag.*, **56**, 321–4.

Martin, J. and Escarre, A. (1980). Datos de fenologia, reduccion de hoja y variaciones estacionales de la composicion foliar, en quatro especies del matorral costero del Mediterraneo Meridional Iberico. *Mediterránea*, **4**, 69–88.

Martin, J. A. de J. (1990). *Estudo histórico-monográfico da Freguesia de Ferragudo do concelho de Lagôa*. Algarve em Foco, Faro.

Martins Loução, M. A. and Brito de Carvalho, J. H. (1989). *A cultura da alfarrobeira*. Ministério da Agricultura, Pescas e Alimentação, Lisbon.

Matos, A. T. de (1980). *Transportes e comunicações em Portugal Açores e Madeira (1750–1850)*. Universidade dos Açores, Ponte Delgâda.

Matos Fortuna, E. R. de (1984). Algarve agricola. *3° Congresso do Algarve. Textos das comunicações*, Vol. 2, pp. 991–7. Racal Clube, Montechoro.

Mattos Parreira, J.J. de (1919). *A oliveira no Algarve*. Tipografia Democratica, Tavira.

Meiggs, R. (1982). *Trees and timber in the ancient Mediterranean world*. Clarendon Press, Oxford.

Meith, C., Micke, W. C., Rough, D., Rizzi, A. D., and Teviotdale, B. (1983). *Almond production*. Division of Agricultural Sciences, University of California, Berkeley.

Mendóça Frazão, M. de (1992). O megasismo de 1755 no Algarve. *7° Congresso do Algarve. Textos das comunicações:* 31–44. Racal Clube, Vilamoura.

Mendonça Pinto, F. (1984). *O Algarve no contexto nacional: situação regional e estratégia de actuação*. Commissão de Coordenação da Região do Algarve, Faro.

Michael, P. W. (1964). The identity and origin of varieties of *Oxalis pes-caprae* L. naturalized in Australia. *Trans. R. Soc. S. Austr.*, **88**, 167–73.

Mimoso Henriques Cunha, A. P. (1989). Contribuição para o conhecimento da flora algológica da Ria Formosa. *Comm. II Congresso de áreas protegidas: Fundação Calouste Gulbenkian, Lisboa, 4–8 de Dezembro de 1989*, pp. 323–9. SNPRCN, Lisbon.

Mimoso Loureiro, J. J. (1988). Régime hidrológico dos cursos de água do Algarve no periodo de 1970–71 a 1984–85 em relação ao periodo total das séries de observações. *5° Congresso do Algarve. Textos das comunicações*, Vol. 1, pp. 623–8. Racal Clube, Silves.

Mimoso Loureiro, J. J. and Fonseca Nunes, M. (1983). *Monografia hidrológica do Algarve*. Universidade do Algarve, Faro.

Mira Galvão, J. (1939). *Manual do podador de oliveiras*. Biblioteca de Agricultura Alentejana. Minerva Comercial, Beja.

Mira Galvão, J. (1940). *O olival*. Junta Nacional do Azeite, Lisbon.

Miranda, J. de (1909). *Cultura da figueira no Algarve*. Instituto Geral das Artes Graphicas, Lisbon.

Modesto, M. de L. (1988). *Cozinha tradicional Portuguesa*. Verbo, Lisbon.

Montalverne, G. (1988). Orquideas: espécies Portuguesas em perigo. *Correio Natur.*, **3**, 27–32.

Monteiro Alves, A. A. (1988). *Técnicas de produção florestal*. Instituto Nacional de Investigação Científica, Lisbon.

Montgomery, K. R. and Strid, T. W. (1976). Regeneration of introduced species of *Cistus* (Cistaceae) after fire in southern California. *Madroño*, **23**, 417–27.

Montoya Oliver, J. M. (1988). *Chopos y choperas*. Ediciones Mundi-Prensa, Madrid.

Montoya Oliver, J. M. (1990). *El pino piñonero*. Ediciones Mundi-Prensa, Madrid.

Morais, A. and Carvalho, C. (1992). A pesca no Algarve: principais números. *7° Congresso do Algarve. Textos das comunicações:* 419–23. Racal Clube, Vilamoura.

Morais Arnaud, J. and Campos, V. (1986). Protecção e conservação sepultos megalíticos do Algarve. *4° Congresso do Algarve. Textos das comunicações*, Vol. 1, pp. 61–8. Racal Clube, Silves.

Moreira, M. da C. (1978). *Apontamentos históricos sobre Castro Marim*. Secretaria de Estado do Ordenamento e Ambiente, Lisbon.

Moreira, M. E. and Oliveira, E. M. (1987). Classificação da cobertura biofísica da Ria Formosa. Seminário 'Programa Corine'. *ICT Informação Técnica: Hidráulica/Detecção*. **ITH 29.** Laboratório Nacional de Engenharia Civil, Lisbon.

Moreira, M. E. S. A., (1989). *Litoral do Algarve*, **ITDR 4**. Laboratório Nacional de Engenharia Civil. Lisbon.

Moreira, M. E. S. A., Coelho, V., Silvestre, S., Dias, C., Oliveira, E. M., and Henriques, R. G. (1989). *Litoral do Algarve: interpretação de imagens digitalizadas obtidas por satélites Landsat*. Laboratório Nacional de Engenharia Civil, Lisbon.

Moreira da Silva, A. & Fos (1982). *Árvores de fruto*. Catálogo. Oporto.

MPAT. (1989) *Relatorio do estado do ambiente e ordenamento do territorio*. Ministério do Planeamento e da Administração do Território — Secretaria de Estado da Administração

Local e do Ordenamento do Território — Secretaria de Estado do Ambiente e dos Recursos Naturais, Lisbon.

Naveh, Z. (1975). The evolutionary significance of fire in the Mediterranean region. *Vegetatio*, **29**, 19–208.

Naveh, Z. and Whittaker, R. H. (1979). Structural and floristic diversity of shrublands and woodlands in northern Israel and other Mediterranean areas. *Vegetatio*, **41**, 171–90.

Neto Martins, A. G. and Branco, R. H. F. (1990). Secagem de uva: comportamento de 7 castas apirénicas (ensaio realizado em 1987). *Experimentação*, **70**. Direcção Regional de Agricultura do Algarve, Faro.

Neves, M. A. dos Ramos das (1985). *A cultura da amendoeira no Algarve*. Instituto Superior de Agronomia, Lisbon.

Noailles, Vicomte de and R. Lancaster (1977). *Mediterranean plants and gardens*. Floraprint, Nottingham.

Nolasco, G. (1988). Tristeza dos citrinos. *5° Congresso do Algarve. Textos das comunicações*, Vol. 2, pp. 901–7. Racal Clube, Silves.

Noonan, L. A. (1989). *John of Empoli and his relations with Afonso de Albuquerque*. Instituto de Investigação Científica Tropical, Lisbon.

Núncio, M. T. G., Ramos, L. and Ramalho, J. (1988). Programa de aproveitamento integrada dos recursos hídricos do Algarve. *5° Congresso do Algarve. Textos das comunicações*, Vol. 1, pp. 617–22. Racal Clube, Silves.

Nunes, A. M. A. (1985). *Alcoutim*. Câmara Municipal do concelho de Alcoutim.

Obenauf, G., Gerdts, M., Leavitt, G., and Crane, J. (1978). *Commercial dried fig production in California*. Division of Agricultural Sciences, University of California, Berkeley.

Oliveira Marques, A. H. de (1978). *Introdução à historia da agricultura em Portugal: a questão cerealifera durante a Idade Média*, 3rd edn. Edições Cosmos, Lisbon.

Oliveira Marques, A. H. de (1985). *História de Portugal I*, 12th edn. Palas Editora, Lisbon.

Oliveira Marques, A. H. de (1987). *A sociedade medieval Portuguesa*, 5th edn. Sá da Costa, Lisbon.

Ornduff, R (1987). Reproductive systems and chromosome races of *Oxalis pes-caprae* L. and their bearing on the genesis of a noxious weed. *Ann. Miss. Bot. Gdn.*, **74**, 79–84.

Orwell, G. (1938). *Homage to Catalonia*, Secker and Warburg, London.

Owen, J. (1980). *Feeding strategy*. Oxford University Press.

Paglietta, R. (1986). *El frambueso*. Ediciones Mundi-Prensa, Madrid.

Palhavã, M. (1988*a*). Algarve: novas hipóteses frutícolas. *5° Congresso do Algarve. Textos das comunicações*, Vol. 2, pp. 879–87. Racal Clube, Silves.

Palhavã, M. (1988*b*). Amendoeiras e pessegueiros para o Algarve. *5° Congresso do Algarve. Textos das comunicações*, Vol. 2, pp. 915–23.

Palmeirinha Godinho Rebelo, M. de L. and Couto da Rocha Pita, J. R. (1992). Aplicação dermofarmaceutica da goma de alfarroba em preparações antritranspirantes. *7° Congresso do Algarve. Textos das comunicações*: 353–358. Racal Clube, Vilamoura.

Paquete, M. (1988). *Algarve: guia histórico turístico*. Lixbuna Edições, Amadora.

Parker, P. F. (1981). The endemic plants of metropolitan Portugal. A survey. *Bol. Soc. Brot. II*, **53**, 943–94.

Parker, S. and Stevens Cox, G. (1970). *The giant cabbage of the Channel Islands*. Toucan Press, Guernsey.

Parsons, J. J. (1962). The cork oak forests and the evolution of the cork industry in southern Spain and Portugal. *Econ. Geogr.*, **38**, 195–214.

Passos de Carvalho, J. (1986). Fitotécnica, fenologia, pragas dos citrinos e protecção integrada. *4° Congresso do Algarve. Textos das comunicações*, Vol. 2, pp. 1019–24. Racal Clube, Silves.

Pato, O. (1988). *O vinho: sua preparação e conservação*, 8th edn. Clássica Editora, Lisbon.

Payne, S. (1968). The origins of domestic sheep and goats: a reconsideration in the light of the fossil evidence. *Proc. Prehist. Soc.*, **34**, 368–84.

Pemberton, R. W., and Turner, C. E. (1989). Occurrence of predatory and frugivorous mites in leaf domatia. *Am. J. Bot.*, **76**, 105–12.

Pena, A., Gomes, L., and Cabral, J. (1985). *Fauna e flora de Mértola.* Câmara Municipal de Mértola.

Pereira, H., Rosa, M. E., and Fortes, M. A. (1987). The cellular structure of cork from *Quercus suber. IAWA Bull.*, **8**, 213–18.

Pereira de Mattos, J. F. (1900). *Cultura da alfarrobeira.* Instituto d'Agronomia e Veterinaria, Lisbon.

Perez Bueno, M. (1989). *El azafran.* Agrogúias Mundi-Prensa, Madrid.

Pessoa, F. (1988*a*). *Contribuição para a estratégia nacional de conservação.* Serviço Nacional de Parques, Reservas e Conservação da Natureza, Lisbon.

Pessoa, F. (1988*b*). Depois do Ano Europeu do Ambiente. *Correio Natur.*, **3**, 6–9.

Piggott, S. (1965). *Ancient Europe.* Edinburgh University Press.

Pinto, M. M. (1987). *Santa Bárbara de Nexe.* Estudo monográfico, Santa Bárbara de Nexe.

Pinto da Silva, A. R. (1963). L'étude de la flore vasculaire du Portugal Continental et des Açores les dernières années (1955–1961). *Webbia*, **8**, 397–413.

Pinto Peixoto, J. (1987). *O homen, o clima e o ambiente*, 3 volumes. Ministério do Planeamento e da Administração do Território — Secretária de Estado do Ambiente e dos Recursos Naturais, Lisbon.

Placito, P. J. (1986). *Iris planifolia. Iris Year Book 1986*, pp. 65–66. British Iris Society, Bridgwater.

Placito, P. J. (1987). *Gynandriris sisyrinchium* in the Algarve. *Iris Year Book 1987*, pp. 72–3. British Iris Society, Bridgwater.

Placito, P. J. (1991). The reptiles of the Algarve. *Algarve Mag.*, **154**, 44–7.

Polunin, O. (1969). *Flowers of Europe: A field guide.* Oxford University Press.

Polunin, O. and Everard, B. (1976). *Trees and bushes of Europe.* Oxford University Press.

Polunin, O. and Smythies, B. E. (1973). *Flowers of south-west Europe.* A field guide. Oxford University Press.

Polunin, O. and Walters, M. (1985). *A guide to the vegetation of Britain and Europe.* Oxford University Press.

Pons, A. and Quézel, P. (1985). The history of the flora and vegetation and past and present human disturbance in the Mediterranean region. In *Plant conservation in the Mediterranean area* (*Geobotany* **7**, ed. C. Gómez-Campo), pp. 25–43. Junk, Dordrecht.

Poore, M. E. D. and Fries, C. (1985). *The ecological effects of eucalyptus.* Food and Agriculture Organization of the United Nations, Rome.

Preston C. D. and Sell, P. D. (1988). The Aizoaceae naturalized in the British Isles. *Watsonia*, **17**, 217–45.

Proctor, M. and Yeo, P. (1973). *The pollination of flowers.* Collins, London.

Pullan, R. A. (1988*a*). Salinas in the western Algarve. *A Rocha Observatory Rep.*, **1988**, 26–31.

Pullan, R. A. (1988*b*). An introduction to the salinas of the western Algarve. *A Rocha Observatory Rep.*, **1988**, 32–6.

Pullan, R. A. (1988*c*). A survey of the past and present wetlands of the western Algarve. *Liverpool Pap. Geog.*, **2**. University of Liverpool.

Purseglove, J. W. (1968–72). *Tropical crops*, 4 volumes. Longman, London.

Quéro, J. C., Hureau, J. C., Karrer, C., Post, A., and Saldanha, L. (1990). *Clofeta. Check-list of the fishes of the eastern tropical Atlantic*, 3 volumes. Junta Nacional de Investigação Científica e Tecnológia, Lisbon.

Rabaça, J. E., (1988). *Estudo sobre os hábitos alimentares da população Cegonha-Branca* (*Ciconia ciconia* L.) *nidificante do Alto Alentejo (Évora-Portugal).* Serviço Nacional de Parques, Reservas e Conservação da Natureza, Lisbon.

Ragazini, D. (1985). *El kaki.* Ediciones Mundi-Prensa, Madrid.

Ramos, D. E. (1985). *Walnut orchard management.* University of California, Davis, California.

Ramos, L., Nuncio, T., Borralho, M. E., Pais, J. R., and Vlachos, E. (1988). *Os recursos hídricos no sul de Portugal,* Vol. II. Ministério do Planeamento e da Administração do Territorio, Lisbon.

Raposo, L. (1986). A pré-historia antiga na costa sudoeste. pp. 75–82. In Liga para a Protecção da Natureza, *Ambiente em discussão,* Lisbon.

Rau, V. (1984). *Estudos sobre a história do sal Português.* Editorial Presença, Lisbon.

Rebelo, T. and Rito, P. (n.d.). *A floresta,* pp. 6. Serviço Nacional de Parques, Reservas e Conservaçao da Natureza, Lisbon.

Reche Marmol, J. (1988). *La sandia,* 3rd edn. Mundi-Prensa and Ministério de Agricultura Pesca y Alimentacion, Madrid.

Rego, F., Botelho, H., and Marques, L. F. (1989). O uso do fogo controlado nos ecossistemas da Serra da Malcata. *Comm. II Congresso de áreas protegidas: Fundação Calouste Gulbenkian, Lisboa, 4–8 de Dezembro de 1989,* pp. 521–30. SNPRCN, Lisbon.

Reimão, D. and Nunes, L. (1989). *Um estudo sobre a impregnabilidade de madeiras redondas de Eucalipto comun.* INCES 7. Laboratório Nacional de Engenharia Civil, Lisbon.

Reis Cunha, F. (1957). *O clima do Algarve.* Universidade Técnica de Lisboa (Instituto Superior de Agronomia), Lisbon.

Reis Moura, A. (1989). Subsidio para a conhecimento da distribuição geográfica actual do camaleão (*Chamaeleo chamaeleon* L.). *Comm. II Congresso de áreas protegidas: Fundação Calouste Gulbenkian, Lisboa, 4–8 de Dezembro de 1989,* pp. 675–80. SNPRCN, Lisbon.

Reuther, W., Batchelor, D., and Webber, H. J. (1967–89). *The citrus industry,* 5 volumes. University of California, Berkeley, California.

Ribeiro, O. (1980). Le caroubier, ses conditions naturelles, son expansion, ses rapports avec l'agriculture. *Portug. Acta Biol. A,* **16,** 3–10.

Ribeiro, O., Lautensach, H., and Daveau, S. (1987–9). *Geográfia de Portugal: I* (1987), *A posição geográfica e o território; II* (1988), *O ritmo climático e a paisagem; III* (1989) *O povo Português.* Sá da Costa, Lisbon.

Ripado, M. F. B. (n.d.). *A cultura da cebola.* Livraria Popular de Francisco Franco, Lisbon.

Rivera Nuñez, D., and Walker, M. J. (1989). A review of the palaeobotanical findings of early *Vitis* in the Mediterranean and of the origins of cultivated grape-vines, with special reference to new pointers to prehistoric exploitation in the western Mediterranean. *Rev. Palaeobot. Palynol.,* **61,** 205–37.

Robinson, J. (1986). *Vines, grapes and wines.* Mitchell Beazley, London.

Rocha Afonso, M. da L. and McMurtrie, M. (1991). *Plantas do Algarve.* SNPRCN, Lisbon.

Rocha Faria, J. M., Godinho, S., Almeida, M. J. R., and Machado, M. S. (1981). *O clima de Portugal. Fasc. XXVII. Estudo hidroclimatológico da Região do Algarve.* Instituto Nacional de Meteorologica e Geofísica. Lisbon.

Rodriguez de los Santos, M., Cuadrado, M., and Arjona, S. (1986). Variation in the abundance of blackcaps (*Sylvia atricapilla*) wintering in an olive (*Olea europaea*) orchard in southern Spain. *Bird Study,* **33,** 81–6.

Romariz, C. and Andrade, C. (1989). Valorização de monumentos geológicos I: O Litoral Sul de Algarve. *Comm. II Congresso de áreas protegidas: Fundação Calouste Gulbenkian, Lisboa, 4–8 de Dezembro de 1989,* pp. 909–15. SNPRCN, Lisbon.

Romero Magalhães, J. (1970). *Algarve económico durante o Século XVI.* Edições Cosmos, Lisbon.

Romero Magalhães, J. (1988). *O Algarve económico 1600–1773.* Imprensa Universitária, Lisbon.

Rosa Santos, L. F. (1989). *Pesca do atum no Algarve.* Parque Nacional da Ria Formosa, Loulé.

Rothmaler, W. (1943). Promontorium Sacrum. Vegetationsstudien im südwestlichen Portugal. I Teil. Die Pflanzengesellschaften. *Fedde's Repert. Beih.,* **128**.

Ruano, M. T., Bastos, M. G., Silva, M. C., and Nascimento, P. (1987). *Ria Formosa: Subsídos para a bibliografia do litoral do sotavento Algarvio.* Centro Estudos Ciências da Terra — Serviço Nacional de Parques, Reservas e Conservação da Natureza, Lisbon.

Rufino, R. (1989). *Atlas das aves que nidificam em Portugal Continental.* Secretaria de Estado do Ambiente e dos Recursos Naturais, Lisbon.

Salvador, J. A. and Ramos, L. (1989). *O livro dos vinhos.* Editorial Fragmentos, Lisbon.

Samson, J. A. (1980). *Tropical fruits.* Longman, London.

Samways, M. J. (1983). Community structure of ants (*Hymenoptera: Formicidae*) in a series of habitats associated with *Citrus. J. App. Ecol.,* **20**, 833–47.

Santos, J. L. dos (1984). Que estrutura agrícola para o minifúndio Algarvio? *3° Congresso do Algarve. Textos das comunicações,* Vol. 2, pp. 983–89. Racal Clube, Montechoro.

Santos Nascimento, B. dos (1944). *A cultura da batata doce na província do Algarve.* Universidade Técnica de Lisboa (Instituto Superior de Agronomia), Lisbon.

Schönfelder, I. and Schönfelder P. (1990). *Wild flowers of the Mediterranean.* Collins, London.

Secretaria de Estado do Ambiente (1988*a*). *Mapas. Ordenamento Biofísico do Litoral Algarve. Espécies características do Algarve. P 12 (2 mapas), Escala 1:100 000.* Secretaria de Estado do Ambiente, Lisbon.

Secretaria de Estado do Ambiente (1988*b*). *Mapas. Ordenamento Biofísico do Litoral Algarve. Espécies não endémicas características do Algarve. P 13 (2 mapas).* Secretaria de Estado do Ambiente, Lisbon.

Secretaria de Estado do Ambiente (1988*c*). *Mapas. Ordenamento Biofísico do Litoral Algarve. Endemismos Portugueses existentes no Algarve. P 14 (2 mapas).* Secretaria de Estado do Ambiente, Lisbon.

Secretaria de Estado do Ambiente (1988*d*). *Mapas. Ordenamento Biofísico do Litoral Algarve. Espécies raras da flora Portuguesa. P 15 (2 mapas).* Secretaria de Estado do Ambiente, Lisbon.

Seigue, A. (1985). *La forêt circum-méditerranéenne et des problèmes.* Maisonneuve et Larose, Paris.

Sérgio, C., Sim-Sim, M., Casas, C., Cros, R. M., and Brugués, M. (1984). A vegetação briológica das formações calcárias de Portugal — II. O barrocal Algarvio e o Promontório Sacro. *Bol. Soc. Brot. II,* **57**, 275–307.

Serrano, A. R. M. (1988). *Contribuição para o conhecimento dos coleópteros da Reserva Natural do Sapal de Castro Marim — Vila Real de Santo António. Os carabídeos. (Coleoptera. Carabidae).* Serviço Nacional de Parques, Reservas e Conservação da Natureza, Lisbon.

Serrão, J. (1986) *Cronologia geral da história de Portugal.* Livros Horizonte, Lisbon.

Service, N. (1989). *Iris planifolia* in the wild. *Spectrum,* **2**, 4. New Zealand Iris Society, Tauranga.

Serviços Geológicos de Portugal (1970). *Carta das nascentes minerais de Portugal.* Lisbon.

Silva Carvalho, A. da (1939). *Memórias das Caldas de Monchique.* Comissão Administrativa das Caldas de Monchique.

Silva Ferreira Sampaio, G. A. da. (1949). *Iconografia selecta da flora Portuguesa.* Ministério da Educação Nacional. Instituto para a Alta Cultura, Lisbon.

Silva Ferreira Sampaio, G. A. da (1988). *Flora Portuguesa,* 4th edn. (facsimile). Instituto Nacional de Investigação Científica, Lisbon.

Silva Lopes, J. B. da (1988). *Corografia ou memória económica, estatística e topográfica do Reino do Algarve,* 2 volumes. Algarve em Foco, Faro.

Silveira da Costa, M. A. (1985). *Árvores e arbustos florestais: árvores florestais (resinosas),* Vol. II. Editora Livraria Popular Francisco Franco, Lisbon.

Simmonds, N. W. (1976). *Evolution of crop plants*. Longman, London.

Simmons, A. F. (1972). *Growing unusual fruit*. David and Charles, Newton Abbot.

SNPRCN (1986). *Ria Formosa, planeamento*. Serviço Nacional de Parques, Reservas e Conservação da Natureza, Lisbon.

Soares, J. M. (1965). *Os frutos e produtos hortícolas na economia do Algarve*. Junta Nacional das Frutos, Lisbon.

Soares Barreto, L. (1988). *A floresta: estrutura e funcionamento*. Serviço Nacional de Parques, Reservas e Conservação da Natureza, Lisbon.

Soares Chaves, J. A. (1980). *A protecção dos citrinos*. Direcção Regional do Agricultura do Algarve, Lisbon.

Soares Chaves, J. A. (1988). *Inimigos das culturas*. Ministério da Agricultura Pescas e Alimentaçaão, Lisbon.

Sousa Pontes, A. de (1982). A colheita mecânica das azeitonas e dos frutos secos do Algarve. *2° Congresso Nacional sobre o Algarve. Textos das comunicações*, pp. 571–5, Racal Clube, Balaia.

Sousa Veloso, J. C. S. de Garrido, J., and Bettencourt, J. M. (1988). *Horticultura e floricultura*, 2nd edn. Editorial Notícias, Lisbon.

Spina, P. (1989a). *El pistacho*. Ediciones Mundi-Prensa, Madrid.

Spina, P. (1989b). *El algarrobo*. Agroguías Mundi-Prensa, Madrid.

Stanislawski, O. (1962). The Monchique of southern Portugal. *Geogr. Rev.*, **52**, 37–55.

Stevenson, A. C. (1985). Studies in the vegetational history of S. W. Spain. II. Palynological investigations at Laguna de las Madres, S. W. Spain. *J. Biogeogr.*, **12**, 293–314.

Stowell, P. (1989). Boat-building is getting harder to find. *Algarve Mag.*, March 1990, 48–51.

Strasburger, E. (1906). *Rambles on the Riviera* (Trans. O. and B. Comerford Casey). Unwin, London.

Sutton, D. A. (1988). *A revision of the tribe Antirrhineae*. British Museum (Natural History) and Oxford University Press.

Taborda de Morais, A. (1938). Breves estudos na flora Portuguesa. *Ann. Soc. Brot.*, **41**, 25–36.

Tamaro, D. (1984). *Fruticultura*, 4th edn. Gustavo Gili, Barcelona.

Tavares, C. N. (1961). Perigos da introdução de vegetais exóticos. *Prot. Natur.*, n.s. **5–6**, 13–20.

Tavares, C. N. and Saccarão, G. F. (1960). A protecção à natureza em Sagres–S. Vicente. Seu interesse e urgência. *Prot. Natur.*, n. s. **3–4**, 1–18. (See also **5–6**, 25, 1961).

Taylor, A. R. (1977). Lightning and trees. *Lightning* (ed. R. H. Golde), Vol. 2, pp. 831–49. Academic Press. London.

Taylor, A. W. (1972). *Wild flowers of Spain and Portugal*. Chatto and Windus, London.

Taylor, G. W. (1990). Ancient textile dyes. *Chem. Br.*, **26**, 1155–8.

Teixera Gomes, P. (1986). *Mamíferos do Parque Nacional Peneda-Gerês: I-O rato-do-campo (Apodemus sylvaticus L., 1758)*. Serviço Nacional de Parques, Reservas e Conservação da Natureza, Lisbon.

Tello Queiroz J. (1992). Da importância de Lagos nos Descobrimentos Henriquinos. *7° Congresso do Algarve. Textos das comunicações*, pp. 65–71. Racal Clube, Vilamoura.

Thomas, D. K. (1979). Figs as a food source of migrating garden warblers in southern Portugal. *Bird Study*, **26**, 189–91.

Tous Marti, J. and Batlle Caravaca, I. (1990). *El algarrobo*. Ediciones Mundi-Prensa, Madrid.

Trabaud, L. (1977). Comparison between the effect of prescribed fires and wood fires on the global quantitative evolution of the Kermes scrub oak (*Quercus coccifera* L.) garrigues. In *Proceedings of the symposium on environmental consequences of fire and fuel management in Mediterranean ecosystems* (ed. H. A. Mooney and C. E. Conrad), pp. 271–82. Forest Service, US Dept. Agriculture, Washington, DC.

Trabaud, L. (1982). Effects of past and present fire on the vegetation of the French Mediterranean region. In *Proceedings of the symposium on dynamics and management of*

Mediterranean-type ecosystems, (ed. C. E. Conrad and W. C. Occhel), pp. 450 7. Pacific Southwest Forest and Range Experimental Station, Berkeley, California.

Trabaud, L. (1983). The effects of different fire régimes on soil nutrient levels in *Quercus coccifera* garrigue. *Ecol. Stud.*, **43**, 233–43.

Trabaud, L. (1987). Fire and survival traits of plants. In *The role of fire in ecological systems* (ed. L. Trabaud), pp. 63–89. SPB Publishing, The Hague.

Trabaud, L. and de Chanterac, B. (1985). The influence of fire in the phenological behaviour of Mediterranean plant species in Bas-Languedoc (southern France). *Vegetatio*, **60**, 119–30.

Trabaud, L. and J. Lepart (1980). Diversity and stability in garrigue ecosystems after fire. *Vegetatio*, **43**, 49–57.

Trancoso Vaz, J. (1955). *Uva de mesa*. Livraria Luso-Espanhola, Lisbon.

Trancoso Vaz, J. (1987*a*). *Notas sobre frutos secos*. Ministério da Agricultura, Pescas e Alimentação, Lisbon.

Trancoso Vaz, J. (1987*b*). *Uvas de mesa cultivadas em Portugal*. Direcção-Geral de Planeamento e Agricultura, Lisbon.

Tutin, T. G., Heywood, V. H., Burges, N. A., Valentine, D. H., Walters, S. M., and Webb, D. A., (1964–80). *Flora Europaea*, 5 volumes. Cambridge University Press.

Unwin, P. T. H., (1991). *Wine and the vine. An historical geography of viticulture and the wine trade*. Routledge, London.

Valdés, B., Talavera, S., and Fernandez-Galiano, E. (eds., 1987). *Flora Vascular de Andalucía Occidental*, 3 volumes. Ketres, Barcelona.

Vasconçelos Lima, A. and Gomes de Castro, A. (1988). *Ciência e tecnologia dos materiais*. Universidade do Trás-os-Montes e Alto Douro, Gondomar.

Vedel, H. (1978). *Trees and shrubs of the Mediterranean*. Penguin, Harmondsworth.

Veiga de Oliveira, E., Galhano, F., and Pereira, B. (1983). *Alfaia agrícola Portuguesa*, 2° edn. Instituto Nacional Investigação Científica, Lisbon.

Veiga Fereira, O. da and Leitão, M. (n.d.). *Portugal pré-histórico*, 2nd edn. Europa-Americana, Mira-Sintra.

Vianna e Silva, M. (1983). *A cultura do arroz*, 2nd edn. Clássica Editora, Lisbon.

Viegas Guerreiro, M. and Romero Magalhâes, J. A. (1983). *Duas descrições do Algarve do Século XVI*. Sá da Costa, Lisbon.

Vilhena Mesquita, J. C. (1986). *A viagem: uma outro forma de turismo na perspectiva do conhecimento histórico*. Universidade do Algarve, Faro.

Vitoriano, J. (1980). Agricultura no Algarve. *1° Congresso Nacional sobre o Algarve. Textos das comunicações*. Racal Clube, Aldeia das Açoteias.

Vuillemin, J. and Bulard, C. (1981). Ecophysiologie de la germination de *Cistus albidus* L. et *Cistus monspeliensis* L. *Natur. Monsp., Bot.* **46**, 1–11.

Warburg, O. E. (1931). *Cistus* hybrids. *J. R. Hort. Soc.*, **56**, 217–24.

White, F. (1983). *The vegetation of Africa*. UNESCO, Paris.

Winer, N. (1980). The potential of the carob (*Ceratonia siliqua*). *Int. Tree Crops J.*, **1**, 15–26.

Wojterski, T. W. (1990). Degradation stages of the oak forests in the area of Algiers. *Vegetatio*, **87**, 135–43.

Woodell, S. R. J. (1989). Cape St. Vincent and the Sagres Peninsula, Portugal: important biological sites under threat. *Envir. Cons.*, **16**, 33–9.

Woodell, S. R. J. and McDonald, A. W. (1989). The vegetation of Cape St. Vincent, Algarve, Portugal, *Brit. Ecol. Soc. Bull.*, **20**, 118–20.

Wrigley, J. W. and Fagg M. (1983). *Australian native plants*, 2nd edn. Collins, Sydney.

Wuerpel, C. E. (1974). *The Algarve, Province of Portugal: Europe's southwest corner*. David & Charles, Newton Abbot.

Zapata, M., Cabrera, P., Bañon, S., and Roth, P. (1989). *El melon*. Ediciones Mundi-Prensa, Madrid.

Zedler, P. H. and Scheid, G. A. (1988). Invasion of *Carpobrotus edulis* and *Salix lasiolepis* after fire in a coastal chaparral site in Santa Barbara County, California. *Madroño*, **35**, 196–201.

Zohary, M. (1982). *Plants of the Bible*. Cambridge University Press.

Zohary, D. and Hopf, M. (1988). *Domestication of plants in the Old World*. Oxford University Press.

Zuccherelli, G. and Zuccherelli, G. (1990). *La actinidia (kiwi)*. Ediciones Mundi-Prensa, Madrid.

Index of scientific names

*Page references relating to figures are underlined. References to plates are set in **boldface** type.*

Subject index

Page references relating to figures are underlined. References to plates are set in **boldface** *type.*